**ACPL ITEM
DISCARDED**

No. 640
$7.95

Selecting & Improving Your Hi-Fi System

By Harvey F. Swearer

Blue Ridge Summit, Pa. 17214

Allen County Public Library
Ft. Wayne, Indiana

FIRST EDITION

FIRST PRINTING—FEBRUARY 1973

Copyright ©1973 by TAB BOOKS

Printed in the United States
of America

Reproduction or publication of the content in any manner, without express permission of the publisher, is prohibited. No liability is assumed with respect to the use of the information herein.

Hardbound Edition: International Standard Book No. 0-8306-2640-9

Paperbound Edition: International Standard Book No. 0-8306-1640-3

Library of Congress Card Number: 72-87459

7110821

Preface

Never before has the hi-fi buff had such a wide range of selection as is currently offered in the exciting field of remarkable sound reproduction. He steps inside an ever-expanding circle of sound where he may not only expect more for his dollar, but rest assured that he will have his fondest hopes fulfilled. The author will explain in layman's terms, what the features of each unit in the high fidelity system should be, and what the buyer should look for.

He will not be able to tell you the best system for you to buy, simply because the best system for you is the one that fills your specific needs. Set the price limit and then start looking around: compare prices, features, claims, while listening closely. You don't necessarily want the one the salesman likes best; after all, he won't be the one to enjoy it. Even your neighbor may be enthusiastic over a particular system, but it could be a nightmare for you. How often friends and relatives ask the TV repairman, "what's the best television for me to buy?". He works on all makes every day and he must know what's best—but he doesn't: the expert, if he's honest with you, only knows what is best for himself, and the same is true of every other person—layman, expert, or in between. When you find a system or component part that you like at a satisfactory price, try it in your home with the understanding that you may return it in the event that it fails to perform as expected. If you are really sincere, most firms will gladly go along with this kind of an arrangement, and after all, that's the only way you can be sure.

Improvements in the hi-fi world are constantly appearing. There is no time like the present for updating or even reappraising your system. Field problems may occur, but most manufacturers, thinking progressively, are preparing to cope with them even before they become complaints. Although many difficulties may arise under actual living room conditions, even the most accurate laboratory tests are often not able to bring them to light.

There are exciting new offerings available to the non-technically inclined. Descriptions, specifications, photos and prices where available will be given to facilitate your selection to fit your individual taste, and hopefully your pocketbook as well.

Harvey F. Swearer

Contents

1 **Program Sources** 7
Major Considerations in Record Players—Tape Player Features—Amplifier Action—Microphone Operation—Evaluating the AM-FM Receiver—Exceed Minimum Performance—Improving Your System With Low-Cost Accessories

2 **AM-FM Receivers & Amplifiers** 25
The Kit Approach—Higher-Priced Extras—The Stereo Amplifier Section—Flexibility, Controls, and Output Requirements

3 **Four Channel Sound** 60
Updating Your Stereo With a Converter—System Wiring—Additional Features in Other Units—Wireless Remote 4-Channel AM-FM Stereo Receiver

4 **Stereo Record Players** 85
Observe Tracking, Anti-Skating, and Other Adjustments—Economy Units—Compact Systems—Turntable Features—New Record Playing System Developments

5 **Tape Recorders** 114
Importance of Mechanism Drive—Cassette, Reel-to-Reel, or Cartridge (8-Track)—Accessibility and Control Features—Level Meters—Battery-AC Operation—Tape Problems and Answers—Simple Cleaning and Degaussing—Pushbutton Value—Direct Recording from AM-FM Receiver

6 **Connectors & Cables** 137
Connecting Accessories—Importance of Connecting Cables—Getting Maximum Output to Speaker Systems—

Basic Terminal Layouts—Installing Extra Speakers—External Level Controls—Plugging in Stereo Headphones—Private Listening Features—Plugs and Jacks—Constant Voltage Lines

7 Hi-Fi Speaker Systems   155

Reflector Speaker System—Air Suspension Speaker System—Power Handling Capability—Allowing a Safety Margin—Power and Its Effect on What You Hear—Getting More "Lows" for Less Money—Proper Use of External Controls—Importance of Frequency Range—Additional Speaker Systems

8 Speaker Enclosures   183

Wall, Closet, and Shelf Enclosures—Assembling Speaker Cabinets for Unusual Balance—Suspension Arrangements—Unique Layouts for Depth Effect—Console Cabinets—Aperiodic Design

9 Overall System Checks and Summary   203

Simple Shortcuts and Adjustments—Stereo Accessories—Definitions Prove the Point—How to Evaluate What You Hear

Glossary of Hi-Fi Terms   211

Index   221

Chapter 1
Program Sources

Varying air pressure results in sound, and as the sound wave rises and falls, the movement is sensed by the ear and heard. Sound is the change or alteration in air pressure. Air acts as the carrier and is modulated by music, voice, or noise. The sound wave rises and falls much like waves in a lake, and has a frequency according to the number of rises and falls per second. The sound wave shown in Fig. 1-1 having a single swing (one rise and fall) in one second would be said to have a frequency of one hertz (Hz). A high note could have 10,000 or more of these in a single second of time. Any sound, regardless of whether it came from music, speech, or even ordinary noise, must disturb the air pressure to set up waves of sound, and the louder that sound, the greater the change in air pressure. Sound waves travel in all directions from their source while rising and falling in a manner similar to the pattern produced when a pebble is tossed in the still water of a pond. So the loudness of the sound determines the amplitude

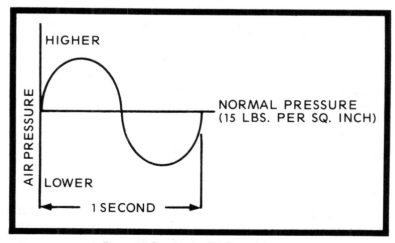

Fig. 1-1. One hertz (Hz) sound wave.

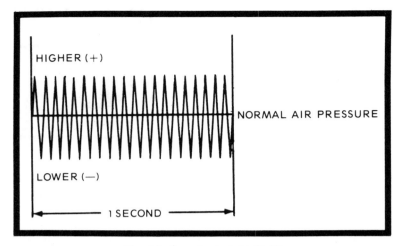

Fig. 1-2. Sound wave at 20 Hz.

(height) of the wave, and the pitch or frequency of the sound corresponds to the number of rises and falls per second. Amplitude is the degree or extent of the swing as measured in decibels (db) and frequency is the number of swings per second measured in hertz (Hz). A low frequency sound wave having a value of 20 Hz is shown in Fig. 1-2.

The pitch of audible sound waves usually covers a frequency range of about 20 through 20,000 Hz, although very few persons are able to hear the high frequency notes above 15,000 Hz. In most cases, the shrill pitch above 15,000 Hz may only be distinguished by the young, while the deep bass notes below 50 Hz may be enjoyed by nearly all. A note in the middle of the audible range could be displayed by the piano at 2,000 Hz, but harmonics (integral multiples) would necessarily be produced to about the fourth harmonic (8,000 Hz) in order to afford depth and clarity to the tone. The amplifier in our system would be required to faithfully amplify these harmonics to ensure high fidelity sound. Otherwise, distortion would result and the overall system would be degraded to a degree that could not be tolerated.

MAJOR CONSIDERATIONS IN RECORD PLAYERS

The record player or automatic turntable is an important source of high quality reproduction in any hi-fi system, and should be very judiciously selected with a good look to the future. It is a primary source of program material for your system and the output can never be better than what you put into it. If you must economize here, go all the way—skip it! After all, you will have other sources of material supplied by the AM-FM stereo receiver and tape player, so why waste funds on a cheap phono player just to have a complete system. Take your time on this vital unit, shop around, compare prices, ask questions, and look for specials; but don't jump at these either. Evaluate each and every one carefully. If it is a discontinued model, make sure there wasn't an unusual number of complaints or dissatisfied customers causing the price reduction.

Smooth, constant speed is essential in phono players and even the slightest variation leads to all sorts of problems, so the motor must be of excellent quality to ensure adequate performance in this area. The synchronous motors used in the better changers are not affected by variations in the usual household AC voltages as they are locked to the frequency of the household supply which is normally very close to 60 Hz at all times. Line voltages do change considerably according to demand; but even during very large changes in voltage, the frequency remains constant at 60 Hz. In addition to the smooth speed of the synchronous motor, multipole arrangements are often designed in the higher quality changers to add power to the smooth rotation, and thus guarantee that that important feature will remain constant. So inspect the motor specifications first. This will enable you to determine whether or not to continue. If they do not measure up on this initial point, by all means—pass it up. A top quality motor could guarantee plus or minus one tenth of one percent speed. While it is not necessary to require this precision unless you want it, use the figure as a reference and come as close as your budget will allow. Actually the figure means a maximum variation in speed from + to — of less than one twentieth of a revolution per minute at thirty three and one-third rpm.

Tone arm tracking force should be less than the optimum values recommended for the better cartridges available, and several of the features should include adjustments available for specific conditions involving cartridge changes, antiskating, and cueing. Whether you already have an extensive record collection or are assembling one as you go, you will want to take the best possible care of them, and this is just one

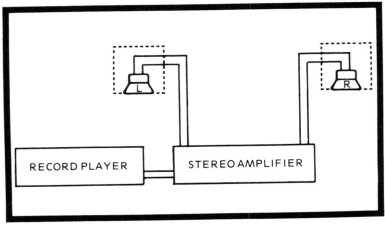

Fig. 1-3. Typical record player system.

of the many features of a good automatic record player. When you play those records, you will enjoy hearing it like it was—not like it is. Look over chapter four before you make your selection or even before you start shopping for possibilities, several of the better turntables are described along with specifications and features provided by the individual manufacturer. There are also many other fine units not mentioned due to lack of space or failure to receive information in time for publication, but the wide price range of listed types will make it easier to compare prices and features of those available to you in your area. Fig. 1-3 shows a typical block diagram of a record player system, and a typical player is pictured in Fig. 1-4 showing one of the better medium priced units by United Audio.

TAPE PLAYER FEATURES

The process of recording on tape requires that a tiny microphone output be stepped-up through a preamplifier before being fed to the recording head. This head contains a coil of wire wrapped around a magnetic core, and as the amplified signals from the microphone flow through the coil, a fluctuating magnetic field is produced at the core gap through which the tape being recorded passes. Iron oxide particles on the tape are arranged according to the magnetic field fluctuations as it passes through the gap of the recording head. The tape is drawn past the head by a revolving roller driven by

Fig. 1-4. Dual model 1215 automatic turntable. (Courtesy United Audio Products, Inc.)

a small induction type motor. To play back the tape, it must be drawn past the head at the same speed as during the recording. During playback, the magnetized particles produce a small magnetic field in the core of the head to generate an electrical wave in its coil. This is exactly the reverse of the recording procedure. The tiny electrical potential is now preamplified before being applied to the main amplifiers as shown in the sketch of Fig. 1-5.

Tape recorders may offer just about as many features and extras as any prospective purchaser could possibly desire and

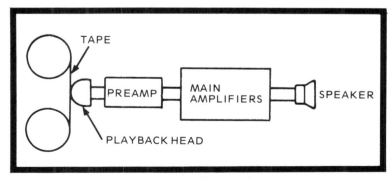

Fig. 1-5. Block diagram of tape playback system.

Fig. 1-6. Sony-Superscope TC-277-4.

more than most would care to buy. The reel to reel type may feature 4 heads (record, playback, and 2 erase); automatic shut-off when tape breaks or comes to the end; front and rear headphone jacks; large twin VU meters; 4 digit index counter with reset button; two or three recording speeds; 4-channel stereo record-playback; hysteresis synchronous motor for constant, smooth power; and wide frequency response. The reel to reel unit probably enjoys the vote of the audiophile and tape enthusiast at this time simply because of the fact that it has a little more to offer in the way of fidelity for making recordings. The cassette and 8-track cartridge player have a large plus in the area of portability, ease of operation, and low cost. However, you will enjoy checking the features of each type as you shop and eventually reach your own correct decision. Fig. 1-6 shows the open reel type.

The cassette tape player is compact and portable, with most offering two- or three-way power operation, which is indeed a worthwhile feature for any who travel. Those who like to work while driving alone on long trips can tape speeches, interviews, and reports with the handy cassette player-recorder. This same unit also provides entertainment or service from pre-recorded tapes, either at home or on the road. The stereo cassette deck may include special heads; brushless motor for direct capstan-drive system; DC motor for fast forward and rewind; feather-touch function control; pause control; noise suppressor; double VU meters; digit tape counter; memory rewind; and auto-stop. Quality would show here in the construction of the motor, heads, and convenient controls that should be fast acting. Needless to say, the tape is extremely critical and lengths over 60 minutes often give trouble. If you wish to record music, buy the better grades of cassette tape and if only speech is recorded do not drop below medium-priced tapes. Fig. 1-7 pictures a popular cassette.

The 8-track stereo cartridge playback deck, popular for some time in cars, is available for home systems to provide stereo playback from the conventional 8-track cartridges or as a quadraphonic source playing 4-channel cartridges. It is definitely a convenient system and offers features including automatic switching between 4-channel and 2-channel stereo; AC motor; slide-in cartridge system; automatic eject pop-up mechanism; push-button function selectors; direct read-out program indicator; and VU meter. It is reasonably priced. Refer to chapter five for additional information on tape players along with specifications of several popular brands.

Questions to Ask—And Answers to Expect

The tape player-recorder segment of the hi-fi system is not only a price versus quality evaluation. First, which of the three types of machines in this group should we consider? If you really like to record, edit, and splice tapes in a professional sort of way, the reel to reel is probably your best bet. However, should you just want good fidelity recording

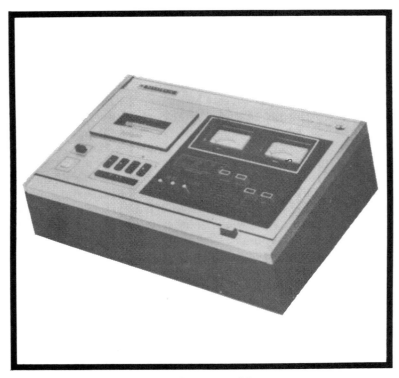

Fig. 1-7. Panasonic RS-275.

with a minimum of fuss and no tinkering, then the cassette is for you. Compact, simple, inexpensive, and with Dolby added, its quality is no less than remarkable. Type number three, the cartridge player (8-track tapes), is mainly for entertainment with pre-recorded tapes. While not as good quality as either of the other two types, it is easy to operate and can supply 4-channel stereo material. So you may wish to consider it as a good source for your future or present 4-channel sound systems; many new tapes are being released by the record companies, but tapes are definitely more expensive. (See Fig. 1-8.)

AMPLIFIER ACTION

The main amplifier must provide an exact replica of the input signal at its output terminals; except that the amplitude or level is thousands of times greater. Any deviation from this carbon copy is correctly termed distortion and must be avoided. The fluctuating (AC) voltage from the microphone, record player, or tape player flows into the amplifer through a capacitor which allows it to pass with very little opposition while blocking the direct current (DC). This audio information reaches the base of the first transistor in our solid-state amplifier, and causes a variation in the control current flowing between the base and emitter. This small current controls a much larger current in the collector circuit of the transistor to produce a magnified replica of the audio information as introduced into the amplifier. It is imperative that our much larger signal be an exact copy of the original except for size or magnitude. Even a tiny variation at this point can be a large change later as the following stages amplify that change along with the original signal. The slightest distortion would then reach objectionable proportions. This is why we carefully select equipment to produce the best fidelity while assembling components in a hi-fi system. Any skimping along the way could result in unacceptable quality from the speaker system. The AR separate amplifier for stereo is shown in Fig. 1-9.

Fig. 1-8. Panasonic 8-track (RS-847US).

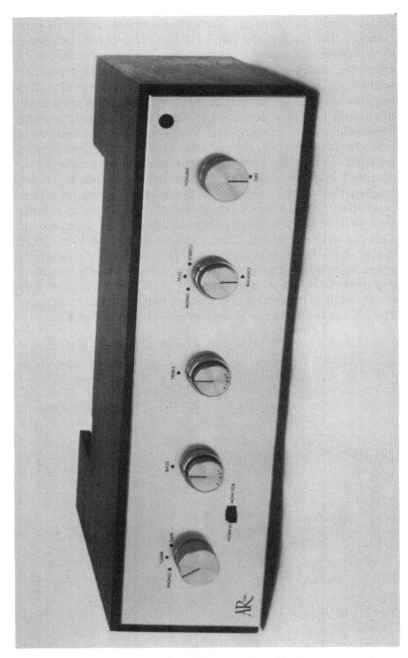

Fig. 1-9. AR stereo amplifier. (Courtesy Acoustic Research, Inc.)

MICROPHONE OPERATION

The microphone converts sound waves to electrical energy and produces a tiny impulse which may be built up to a much larger level with a preamplifier before being fed to the main amplifier. This prevents the tiny impulses from the microphone from being smothered by any interfering noise waves in the area before they can be raised to a high level by the stereo amplifiers. The stereo amplifiers in turn amplify the electrical sound impulses or waves sufficiently to drive the loud speakers at the desired volume. The speaker system promptly converts the higher magnitude audio electronic waves back into sound waves which are carried through the surrounding air to the listener's ear.

Types of Microphones

There are many types of microphones, but all make use of a diaphragm or plate sensitive to the pressure provided by sound waves. The crystal microphone is popular, offers good quality at a low price, but is extremely sensitive to temperature, moisture, and rough handling. It has a frequency response of 50 to 9,000 Hz and usually sells for less than $5. Very similar to the crystal microphone in construction, the ceramic type makes use of ceramic material in place of the crystals for stability. Frequency response is better, 50 to 12,000 Hz, and it is much more stable for almost the same cost. The dynamic microphone uses a diaphragm with a circular coil attached which moves in a magnetic field when actuated. The tiny voltage generated by this action simulates the sound variations. This type features light weight, rugged construction, high output, and excellent frequency response (50 to 15,000 Hz). Cost is higher than other types, but advantages gained are considered to be well worth the difference where quality performance is desired.

EVALUATING THE AM-FM RECEIVER

In the FM section of a receiver, the sensitivity as indicated in microvolts (uv) should be about 3.0 uv or lower (IHF standards). This figure will vary according to price of course, with some of the more expensive receivers under 1.8 uv IHF.

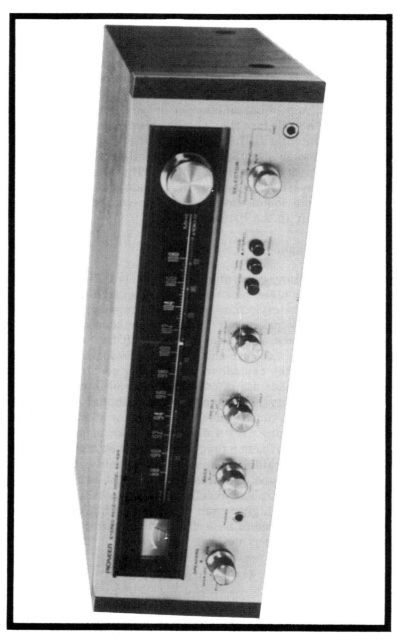

Fig. 1-10. AM-FM stereo receiver model SX-424. (Courtesy U.S. Pioneer Electronics Corp.)

The selectivity rating is even more important in many areas, and is expressed in decibels (db). Here you will notice the less expensive units have about 35 db selectivity while higher priced models show ratings at 100 db or so. The capture ratio, which represents the ability of the tuner to reject unwanted FM stations and interference on the same frequency, should be less than 2.0 db, with a lower figure more desirable. Signal-to-noise ratio (S-N ratio) specifications run close to or about 50 to 75 db with the higher more desirable. Completing the FM considerations from the standpoint of ratings or figures, stereo separation should run 35 db or better and image rejection from 45 db to the excellent 100 db.

The amplifier section of the receiver requires careful consideration as to actual output power desired and the price range. The RMS power in watts per channel should be considered using the same resistance and frequency figures as this will affect the answer considerably. If the amplifier shows a continuous power output of 70 watts per channel at 4 ohms, only 58 watts per channel may be the figure at 8 ohms. So always check specifications to make sure that accurate and fair comparisons are being made. IHF standards of the Institute of High Fidelity Manufacturers may be noted in many of the more important specifications and will be indicated by the designation "IHF." Do not hesitate to ask questions of sales personnel when in doubt. Most will try to be helpful, but be a good listener. Shop around, compare notes outside the store, and by all means do not try to impress with how much you know. It may pay to wait for the "oversell"—often this will lead you to a better buy from a competitor. An AM-FM stereo receiver is pictured in Fig. 1-10.

EXCEED MINIMUM PERFORMANCE

The program source supplying the input to your system, should at least meet average or better requirements. If you later add a better speaker system, your results will be only as good as the source. Always try to bend just a little here at the expense of the part of the system that follows. The output of the best amplifier or the finest speaker system that money can buy, will be no better than the input as provided by the source. After all, this is what high fidelity is all about, getting a realistic reproduction of the original or source material. If you are getting a separate AM-FM tuner, check the sensitivity and selectivity specifications at least. By meeting reasonable minimums here, you will probably be adequately covered in others, and the cost will not be that much greater. The record player is important enough to cover a full chapter in any book—and chapter four goes into the subject from stylus to motor. You may want to use your system for a lot of record-playing enjoyment, which certainly dictates the quality you must select in a phono part, but even though you do not anticipate much use—do not buy any unless it is a good one. The tape player-recorder may be a compromise from a quality angle, this can easily be replaced with a better unit in the future as the need and means coincide. Often included with the tape player-recorder is a good quality microphone and, unless an elaborate recording need is required, this will prove adequate.

IMPROVING YOUR SYSTEM WITH LOW-COST ACCESSORIES

The interference to FM reception may be considerably reduced in most localities. Much improvement to your listening pleasure can be provided by connecting an FM Bandpass Filter at the FM antenna terminals of your receiver. Available at most radio-TV stores for less than $5, this small unit is easily installed with the simple instructions included.

Up to eight times more FM signal is provided with the Winegard FM Supercharger selling for under $15. Easy to install, it uses trouble-free solid-state circuitry and is ideal for FM stereo reception.

A variable high-frequency control is available for matching speaker output to room acoustics, including a control cable for external mounting on the speaker enclosure. Crossover frequency may be set at 2500 or 5000 Hz as desired. The complete kit with instructions costs under $10.

A two- and three-way crossover network with adjustable brilliance and presence can be bought for just under sixteen dollars. Six combinations are possible with this fancy unit: two-way with crossover at 2000 to 3000, or at 5000 Hz; and 3-way with crossovers at 350 and 5000 Hz, at 350 and 2000 to 3000 Hz, at 700 and 2000 to 3000 Hz, and at 700 and 5000 Hz.

The 4-channel stereo adapter will truly improve your 2-channel stereo, and permit you to enjoy those extra channels for less than thirty dollars. There is no second amplifier to buy. Just the adapter and a pair of spare speakers—that you may have stored away in the attic, basement, or garage will provide you a sample of that famous four-dimensional, concert-hall music you have heard so much about. This is derived 4-channel stereo to be sure, but for the small cost involved, the end result will prove most satisfying and you may still add a lot for additional improvement. Just to show how simple this change can be, a block diagram of the converted system is shown in Fig. 1-11, and pictured in Fig. 1-12. Four channel adapters and converters will be reviewed in Chapter 3.

A stereo VU balance meter at $12.95 will make it easy to balance your system as the damped loudness meters offer continuous monitoring of each channel and permit individual attenuation as required.

Matched speaker systems of the three-way type may be used to replace smaller speakers in your system, and will usually ensure considerable improvement in overall performance. The cone tweeter reproduces brilliant highs in systems designed for low and medium efficiency, and works exceptionally well with the low resonance woofer. The linear cones of special design afford good low-end cutoff while showing optimum sensitivity in the high range. The closed

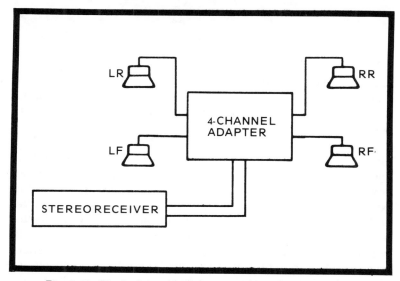

Fig. 1-11. Block diagram of 4-channel adapter connection.

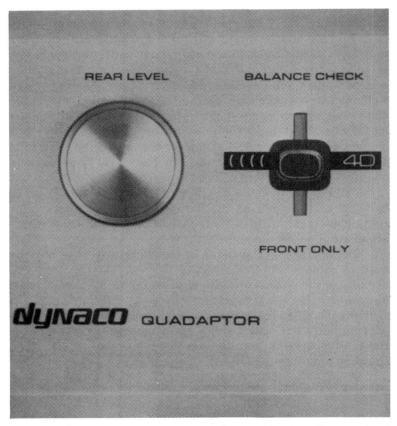

Fig. 1-12. Quadaptor 4-channel stereo adaptor. (Courtesy Dynaco, Inc.)

back on the tweeter eliminates interaction with the mid-range and woofer speakers, while the mid-range horn tweeter almost completely eliminates distortion and introduces amazing results in the important mid and high frequency ranges.

Outdoor speaker system with the Poly-sonic model E-51, all weather design permits use anywhere. Here is a hi-fi speaker at the inexpensive price of only $14.95 each. Peak power is 18 watts, frequency range 70 to 20,000 Hz, input impedance 8 ohms, and available in ivory white, walnut brown, or black. High temperature, high impact styrene enclosure with complementary trim is included, and all necessary hardware for hanging or mounting anywhere—patio, fence, tree, beach, pool, etc. The size is 1⅜" by 6" by 10". They are also available in larger sizes and power and slightly higher in price, but of the same all-weather design. They are bi-directional.

Decorator speakers in attractive paintings or mirrors with Poly-Planar model DS8P or DS8M offer a startling source of hi-fi sound. Art prints in Early American walnut-toned frame have a super-thin speaker behind, completely out of sight. The same unique arrangement is also available with the mirror insert. The speaker has a power handling capacity of 20 watts and a frequency range of 40 to 20,000 Hz. Oval dimensions are 12¼" by 10¾" by 1¼"; impedance is 8 ohms; price is $17.95 each.

The advantage of the Poly-Planar, is its super-thin construction which offers many unique installation possibilities while providing wide-range performance at low distortion. The rugged design meets extremes of shock and vibration, wide temperature environments from minus 20 degrees F to plus 175 degrees F and is moisture-proof. The bi-directional polar pattern with high power capability facilitates additions to present equipment, and the light weight feature permits use in many places previously unavailable. These speakers function in a similar manner to that of the cone type system, with a magnetic structure producing a radial field within a closely spaced gap. As the voice coil is immersed in the gap, signals applied to the coil cause a motor action which produces an acoustical output. Substituting a flat plastic panel for the conventional paper cone reduces the depth requirements to a fraction of that normally needed in the cone structure. The panel is made rigid by the choice of plastic materials in conjunction with a special surface skin treatment. As the material is expanded polystyrene in a compacted bead structure, the beads are largely air and the mass is extremely low. Desired acoustical properties are obtained by circular grooves in conjunction with special acoustic fill materials. This combination provides proper balance with damping for wide response characteristics. A larger piston area is also available with the flat rectangular panel, and therefore length and width dimensions can be made smaller than with the cone speaker, or superior low frequencies can be reproduced for the same equivalent area. The superior rigidity of the acoustic panel eliminates the need for "spiders" or conventional structures to maintain centering of the voice coil in the magnetic gap. The panel is supported directly by a frame of similar polystyrene plastic which provides an identical coefficient of expansion. This results in a stable physical structure substantially independent of wide changes in temperature or ambient conditions. Due to physical characteristics of the panel, the supporting frame may also be shallow without detracting from the rigidity necessary to avoid flexing at low frequencies. Superior mechanical stability permits more accurate coil centering and a shorter magnetic gap along with a lighter magnet for the same power output.

Speaker Operation

Since sound waves result from variations in air pressure, the pump action supplied by the conventional speaker cone appears to be an ideal method of generating movement of air. The speaker cone must move in order to produce sound, so rolls are formed near the edge of the cone to relieve stiffness. Bass tones are produced from big organ pipes, bass fiddles, tubas, and, similarly, from big cones. The size of the speaker cone and the distance it travels are the governing factors in bass reproduction. The 12 and 15 inch cones with soft suspension are fine for bass notes but are far too heavy to vibrate efficiently at high frequencies. Here we need a small, light, stiff cone for the short, fast vibrations at say 10,000 Hz or more. So a high frequency radiator (tweeter) is two or three inches in diameter as opposed to the 12- or 15-inch woofer diameter.

AM-FM Receivers & Amplifiers

Chapter 2

The range in price of FM-AM tuners or FM-AM amplifiers is quite wide depending on the features and extra conveniences offered. In a tuner under $70 we may expect flywheel tuning, VU meter, blackout dial, stereo indicator, less than ½ of 1 percent distortion (0.5 percent harmonic), FM antenna input impedance taps for 75 or 300 ohms, internal AM antenna, 5 microvolts FM sensitivity and a walnut finish cabinet. A similar receiver in kit form could be purchased at a saving of about $25, if you really feel that you want to go to the trouble of careful assembly—some would enjoy this, while others would consider it quite a chore. An amplifier to go along with the tuner would run about the same price as the tuner in either case, and would have an output of possibly 50 watts.

The Lafayette solid-state FM-AM stereo tuner (LT-725A) at less than $130 offers notable features including 1.7 microvolt FM sensitivity, 4 ICs, 2 FETs, 14 transistors, and 19 diodes, 4-gang FM tuning, front and rear panel tape outputs, illuminated signal strength meter, automatic FM stereo switching, stereo indicator, push-button mute switch, built-in FM and AM antennas, AC fuse holder, power-on illuminated dial, 1.5 db capture ratio, and 40 db stereo separation.

Sony offers an outstanding tuner in the ST-5130 which has exceptional sensitivity at 1.5 microvolts and selectivity IHF of 100 db. The FETs insure very low internal noise plus the ability to handle high level FM signals without overloading. It provides outstanding pickup on weak signals, and features muting, AFC, signal strength meter, center station meter, and headphone jack.

The Sony ST-5600 FM stereo or FM-AM tuner is for the listener looking to advancement in his system at a reasonable investment. It offers many of the features found in only the very best equipment. The sensitivity IHF is 3.0 microvolts, selectivity IHF 60 db, plus AFC and signal strength meter.

This unit may be combined with the TA-1010 stereo amplifier to complete a modestly-priced system and includes high filter, loudness contour, speaker selector, AC outlet, and headphone jack. The full complement of functional controls ensure ease of operation under widely varying conditions. The dynamic power output as measured by IHF standards is 44 watts into 8 ohms.

The Lafayette solid-state AM-FM stereo tuner (LT-670) at less than $80 features automatic stereo switching and indicator light, front panel tape output jack, automatic FM station lock (AFC), 30 db stereo separation, multiplex filter, internal FM antenna, 14 transistors, and 16 diodes. Other features include three-gang tuning condenser, 3.5 microvolts FM sensitivity, selectivity 35 db, signal-to-noise ratio 50 db, and image rejection 50 db.

The Eico silicon solid-state FET AM-FM stereo tuner (Model 3300) at $69.95 kit or factory assembled $109.95 has a sensitivity IHF of 3.5 microvolts with a signal-to-noise ratio of 60 db. These specifications are good at this price figure, but do not forget that the kit will require quite a bit of time and work to assemble, and, unless you feel the urge, the factory assembled and tested route may be preferable. The companion stereo amplifier with 50 watts output (IHF 32 watts) is attractively designed, solid-state, and has controls to ensure complete versatility plus the exceptional low-price that matches the tuner.

The Cortina 3780 silicon solid-state FET 50-watt AM-FM stereo receiver, modestly-priced ($109.95 kit, $169,95 wired), is a result of Eico's long experience in electronic design. Advanced RF circuitry is employed using FETs. Silicon solid-state devices ensure temperature stability. Critical circuits are pre-assembled and aligned. A transformer power supply is used for reliability and safety. Styling is attractive, with back-lit panel. FM sensitivity is IHF 3.5 microvolts, signal-to-noise ratio 60 db, image rejection 50 db, and selectivity 20 db. The audio section has an output of 50 watts across 4 ohms (32 watts IHF) with a frequency response of 20 Hz to 20 kHz.

The Pioneer SX-525 ($239.95) AM-FM stereo receiver with 72 watts IHF has an FM sensitivity of 2.2 microvolts IHF. (See Fig. 2-1.) Selectivity is 45 db and the power bandwidth exceeds the usable sound frequency spectrum. The SX-626 model with 110 watts IHF at $279.95 also exceeds the usable sound

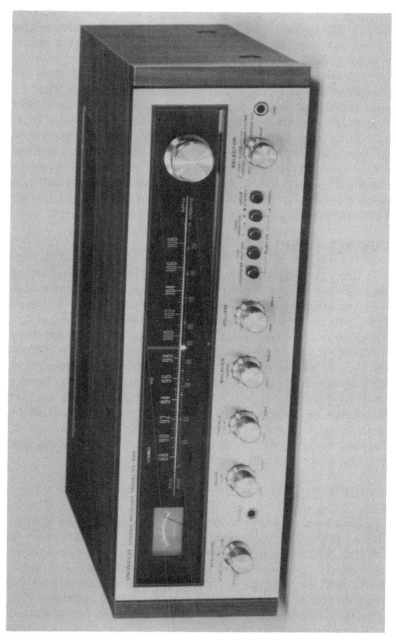

Fig. 2-1. Pioneer model SX-525 AM-FM stereo receiver. (Courtesy U.S. Pioneer Electronics Corp.)

frequency spectrum by a wide margin and offers FM sensitivity IHF as low as 2.0 microvolts plus 70 db selectivity. The SX-727 exhibits 195 watts IHF music power across 4 ohms, sells for $349.95, offers an IHF FM sensitivity of 1.8 microvolts and selectivity of +70 db. The top of the line SX-828 ($429.95) provides 270 watts IHF music power, FM sensitivity IHF 1.7 microvolts, and selectivity of +75 db. The FET (Field Effect Transistor) increases sensitivity, so the SX-828 uses four. Selectivity and capture ratio are optimized by integrated circuits and ceramic filters in the IF stage. Each receiver provides terminals for a wide range of program sources including 4-channel sound. All models have two inputs and two outputs to handle the QL-600 Quadralizer amplifier providing 4-channel sound quite simply by adding two speakers.

THE KIT APPROACH

There is little doubt that a considerable amount of pleasure may be derived from building your very own stereo receiver from one of the many kits available. At the same time, money can be saved as well. However, only you can make that decision—because there will be work and time involved. If you enjoy that type of project, and are confident that you will have the time to carry it through to completion, then by all means give it a try. Otherwise, unless you have a very good friend who will volunteer to help, maybe a smaller kit should be tried first. Of course, all kits can be built by professionals for an extra charge, with the work guaranteed. So, the answer is entirely up to you; think it over carefully and you will make the right decision in your specific case.

HIGHER-PRICED EXTRAS

Sherwood model SEL-300, digital readout stereo FM tuner makes it possible at last to enjoy the computer-accuracy of crystal controlled digital readout of your favorite station, delivered to you with unprecedented sensitivity and selectivity, and the industry's lowest distortion. The set provides pinpoint tuning with two tuning meters and the exclusive FET sideband interchannel hush control, variable for minimum masking of weak stations. This equipment utilizes all-silicon transistors for optimum reliability of performance and carries a three year warranty that includes all parts and labor. Ad-

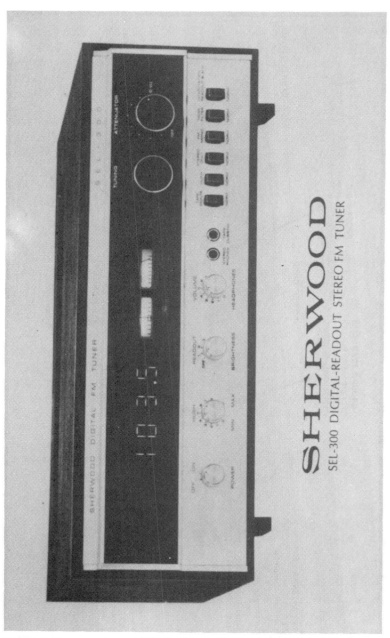

Fig. 2-2. Sherwood model SEL-300 digital-readout stereo FM tuner. (Courtesy Sherwood Electronic Laboratories, Inc.)

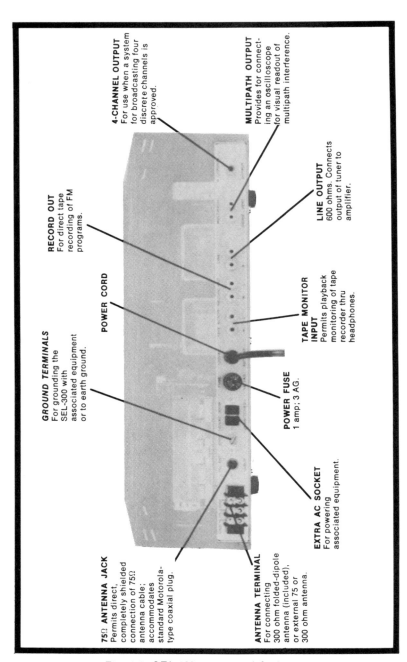

Fig. 2-3. SEL-300 rear panel features.

ditional features include logic circuitry incorporating 17 integrated circuits (ICs), estimated 25-year life on readout tubes, double-sided glass epoxy boards with plated-through interconnections, multi-stage regulation which provides complete immunity to voltage and load variations, 12-pole toroidal IF selectivity-filters (phase-linear, permanently tuned), 3 FET 4-section front end, and a headphone amplifier which provides 300 milliwatts per channel from 20 to 20,000 Hz with less than 0.2 percent distortion. Specifications: sensitivity IHF 1.5 uv; selectivity 85 db; S-N ratio 70 db; harmonic distortion 0.15 percent; image rejection 80 db; stereo separation 40 db; frequency response 20 to 20,000 Hz mono, 20 to 15,000 Hz stereo; dimensions 16¼" W, 5¼" H, 14" D. Power requirements 117V AC, 50-60 Hz, 30 watts. See Fig. 2-2 for pictorial view. Fig. 2-3 displays rear panel layout and features.

Heathkit AJ-1510, digital readout stereo FM tuner is more convenient to tune, easier to read and the computer tuner is the most significant breakthrough in tuner design since solid state. An FM tuner is only as good as its ability to seek, find, and hold broadcast signals as well as its ability to achieve consistently accurate station selection with a minimum of effort. The computer tuner offers incomparable performance in both of these areas. Your station may be programmed with a light touch of the keyboard, no more knob twisting, if it is receivable you can hear it instantly regardless of strong adjacent stations. The slide dial is out, only four softly glowing numerals show where you are—exactly. Touch the reset button and you are clear and ready to program another station. The touch of a button permits you to scan the band as the sweep-scan mode takes over. Computer cards allow you to pre-program your favorite stations. Up to three cards as programmed by you may be inserted in the tuner, and selector buttons A, B, C give you instant access to any of these pre-programmed stations. The end result here is a tuner with specifications you have to hear to believe. Specifications: sensitivity less than 1.8 uV; selectivity greater than 95 db; S-N ratio 65 db; harmonic distortion less than 0.3 percent; image rejection 90 db; stereo separation 40 db; frequency response 20 to 15,000 Hz; dimensions 16⅜" W, 6" H, 14¾" D.

Dynaco model FM-5 transistorized FM tuner is the culmination of years of design research with the objective of

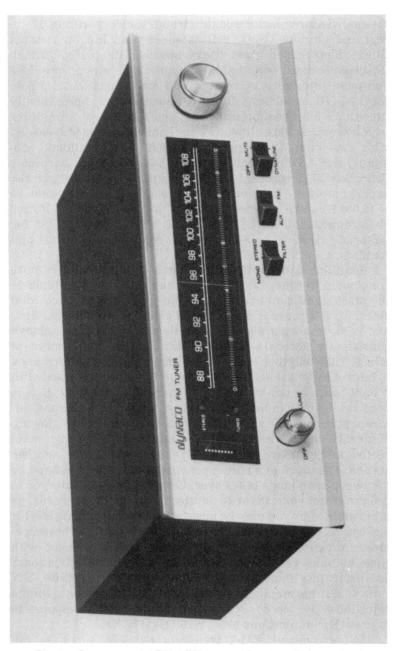

Fig. 2-4. Dynaco model FM-5 FM tuner. (Courtesy Dynaco, Inc.)

offering performance at low cost that is rarely approached by other tuners of the higher price. The circuit automatically fine-tunes a desired station to the minimum distortion point in both mono and stereo, and has a muting circuit that is free of switching transients, regardless of how fast the tuning dial is moved. Specifications include: sensitivity IHF 1.75 uv; selectivity 65 db; S-N ratio 65 db; stereo separation 40 db; harmonic distortion 0.25 percent; capture ratio 1.5 db; dimensions 13½"W, 4¼"H, 9"D. Available in kit form at $149.95 or assembled at $249.95. See pictorial view of the FM-5 in Fig. 2-4.

Harman-Kardon model 820, 140-watt FM stereo receiver is without a doubt one of the best-performing, most versatile stereo receivers designed to sell under three hundred dollars. It has sufficient power to drive four speakers regardless of efficiency or impedance, and is absolutely stable, even when used with full-range electrostatic or low-efficiency speakers. The power amplifier is unlike any found in conventional stereo receivers as it employs the latest low-noise wideband silicon transistors and a heavy duty regulated power supply, enabling the amplifier to extend to below 5 Hz and above 60,000 Hz for flawless reproduction of all harmonics without phase and transient distortion. The output stage ensures accurate balance and symmetry at the clipping points as a high degree of feedback is used to keep distortion low and stability high. Harmonic distortion is below 0.5 percent at full output across the 20 to 20,000 Hz spectrum. Crossover distortion, a phenomenon found in many solid-state stereo receivers, has been reduced to zero. The result is spacious, smooth sound without the stridency that is characteristic of many receivers. Other specifications include FM sensitivity 1.8 microvolts (IHF); image rejection 86 db; stereo separation 35 db; and dimensions of 16⅝" W, 4⅜" H, 12⅞" D.

Panasonic model SA-6500, 200-watt FM-AM stereo receiver features two MOSFETs in the front end, crystal filter, linear FM dial scale, signal strength and center of channel tuning meters, two FM-RF amplifiers, and six FM-IF amplifiers. It has tape monitor, high-low filters, FM muting, MPX high-blend, two phono inputs, speaker selector switch, sliding controls for bass, treble, volume and balance, stereo indicator, push-button operation of all audio switches—and still under four hundred dollars. Specifications include: FM sensitivity 1.8 uv IHF; S-N ratio 60 db; stereo separation 40 db (1 kHz); IHF music power 200 watts (4 ohms); RMS power 70 over 70 (4 ohms); harmonic distortion 0.5 percent (1 kHz, rated output), power bandwidth 5 Hz to 60,000 Hz; dimensions 5⅞"H, 16⅞"W, 15¼"D.

Panasonic model SA-5800, 100-watt FM-AM stereo receiver has two MOSFETs in the front end, and linear FM dial scale. The functional dial pointer brightens when stations are perfectly tuned. There are two tape monitors, two FM RF amps, seven IF amps, and six ceramic FM IF filters. Features include FM muting, center of channel tuning meter, high filter, center channel output, dual bass and treble controls, stereo indicator, and current price of $289.95. Specifications include: FM sensitivity 1.8 uv IHF; S-N ratio 60 db; stereo separation 35 db (1 kHz); music power 100 watts (4 ohms); RMS power 37 over 37 (4 ohms); harmonic distortion 0.5 percent (1 kHz, rated output); power bandwidth 5 Hz to 40,000 Hz; dimensions 5½" H, 16" W, 14" D.

Panasonic model SA-4000, is a 180-watt FM stereo receiver with automatic tuning and remote control. It contains six FETs, two RF stages, a five-section tuning capacitor, six ICs, four ceramic filters, ceramic discriminator, linear FM dial scale, linear sliding controls for tonal quality, computer type push-buttons for function selection, all professional audio controls, and is currently priced at $990.00. Specifications are FM sensitivity 1.5 uv IHF; S-N ratio 70 db; stereo separation 45 db (1 kHz); harmonic distortion 0.1 percent (1 kHz, rated output); power bandwidth 20 Hz to 30,000 Hz; damping factor 150 (8 ohms); IHF music power 180 watts (4 ohms); RMS power 90 over 90 (4 ohms); dimensions 7;;H, 20⅛"W, 17"D.

Sony model STR-6200F FM stereo-FM receiver combines component performance and flexibility features rarely found in even the most costly tuners, preamplifiers, and basic power amplifiers. Twin FETs are used in a common-gate configuration with five tuned circuits for optimum sensitivity and noise-free reception of the weakest, most distant FM signals without adjacent channel interference. Eight solid-state two-element ceramic filters in the IF section replace conventional tuned interstage transformers and provide the ideal text book response curve. Specifications for the FM tuner are sensitivity 1.8 microvolts IHF; image rejection 90 db; IF rejection 100 db; spurious rejection 100 db; AM suppression 65 db; capture ratio

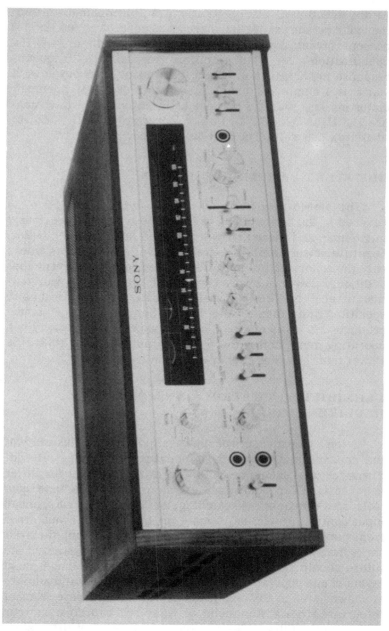

Fig. 2-5. Sony model STR-6200F FM stereo-FM receiver. (Courtesy Sony Corp. of America.)

1.0 db; selectivity 100 db; S-N ratio 70 db; harmonic distortion; mono 0.2 percent at 400 Hz, stereo 0.35 percent at 400 Hz (100 percent modulation); stereo separation 40 db at 400 Hz. Specifications for the amplifier section are frequency response 10-100 kHz + 0, —3 db; continuous RMS power 90-90 watts at 4 ohms, 70-70 watts at 8 ohms (1 kHz); harmonic distortion less than 0.2 percent at rated output. Dimensions are 5⅞"H, 19"W, 15⅞"D. Power source is 117 VAC, 50-60 Hz, at approx. 180 watts. The STR-6200F is shown in Fig. 2-5.

THE STEREO AMPLIFIER SECTION

The stereo receiver includes an amplifier section as covered in the descriptions of the various models previously mentioned, and more that will be listed later in this chapter. Separate power amplifiers are also available in cases where only the tuner section may be selected, or in the event that additional power is desired over that provided by your particular receiver. The solid state amplifier varies according to specific requirements, and in the Sony STR-6200F model produces 360 watts of IHF dynamic power into 4 ohms. When operating into 8 ohms, the power output (IHF) is 245 watts. Power figures will be explained in detail in Chapter 7.

FLEXIBILITY, CONTROLS, AND OUTPUT REQUIREMENTS

Often a true measure of superiority lies in the flexibility of the controls and output-input connection facilities offered. These features should be examined in any receiver-amplifier you consider for your stereo system. Step-type tone controls, speaker equalizer switching, loudness contouring, phono input control, protection circuits for output, numerous other front panel conveniences and ample rear panel input-output jacks for other units that may be added to your system in the future should be kept in mind. Example of the rear panel layout of one popular receiver is shown in Fig. 2-6 and another is shown later in this section. Careful thought here is most important from a standpoint of what you can do with your system in the future. The receiver is the heart of the system and any additions or improvements must start or end here, so

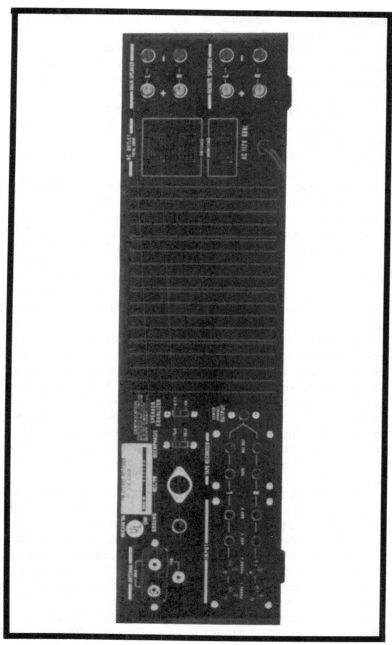

Fig. 2-6. STR-6200F rear panel layout.

stop and consider whether or not you will be able to conveniently accommodate additional equipment.

Additional AM-FM stereo receivers include the following design and specifications as furnished by respective manufacturers. Prices, where listed, are the manufacturer's suggested retail and may vary according to area.

Toshiba model SA-400, AM-FM stereo FM receiver uses integrated circuits (ICs) in almost every section and represents one of Toshiba's most advanced though least expensive receivers. The 40 watts IHF music power results from a unique IC output section that is directly coupled to the speaker system without the usual limitations imposed by output transformers or capacitors. Other features include the multiple frequency response (MFR) tone control circuit; connections provided for all types of systems; and FET FM front end. Specifications are FM sensitivity 2.5 uv IHF; FM selectivity 60 db; stereo separation 35 db; FM image rejection 50 db; output power 40 watts IHF; frequency response 20 to 40,000 Hz; distortion 0.8 percent full power, 0.2 percent at 1 watt; signal-to-noise ratio (S-N) —75 db. Controls are volume; balance; mode; program selector; speaker selector-power switch; tone with variable bass and treble boost-cut plus hinge-point selector switches; tuning. Inputs and outputs are phono aux. inputs (phono jack); tape in and out phone jacks and parallel, single-connection 5-pin DIN jack; main and extension speakers screw terminals; headphone jack; unswitched AC output; 75 and 300 ohm FM antenna connection; inputs and outputs for 4-channel adapter, equalizer, or electronic crossover (phono jacks). Also included are a tuning meter and stereo light indicator. Dimensions are 4½" H, 16½" W, 11¼" D. Weight 14½ lbs.

Toshiba SA-20Y—SA-15Y models represent AM-FM stereo receivers with power to spare. The 20Y offers 120 watts IHF music power, and the 15Y, 100 watts IHF music power. The two models differ only in their power outputs, otherwise, they are identical in all respects. Advanced circuitry enables the FET front end to provide 2.0 microvolts (uv) IHF sensitivity to pick up the weakest stations without overloading on the strongest. Automatic muting eliminates annoying interstation noise. These receivers also feature direct output coupling, and this means no transformers to cause phase shift or rolloff. Other features are the FET tuner with blackout dial, switchable AFC and muting, high and low cut filters, and all-silicon solid-state design. Specifications are FM sensitivity 2.0 uv IHF; FM selectivity 30 db; stereo separation 35 db; FM image rejection 50 db; frequency response 20 to 60,000 Hz; distortion 0.4 percent full power; S-N ratio minus 80 db on aux. Controls are volume (with switchable loudness); bass, treble; AM-FM stereo-phono tape head aux.; mode; mono-reverse-stereo-stereo tape mon-tape L mon-tape R mon; tuning; stereo balance; power off-main speaker extension speaker-headphone switch; switchable AFC, muting, low-cut and high-cut filters. Inputs and outputs are front panel stereo headphone jack; phono, tape head, aux. 1 & aux. 2 inputs; inputs; stereo tape record-playback connections (DIN single-connection and standard US jacks); preamp outputs and main amplifier inputs with external jumper cables (suitable for SC-410 Quad Matrix Converter-Amplifier); 300-ohm FM antenna input; AM antenna input. Also included are tuning meter, FM-stereo light, and protection circuit light. Dimensions are 7¼"H, 19½"W, 15⅜"D. Weight 33 lbs. Power required is 120V, 60 Hz.

Harman-Kardon model 230A, 45 watt AM-stereo FM solid-state receiver exceeds every specification ever established for medium-priced stereo components at a price under $160.00. Frequency response, often limited to the narrow range of 20 to 20,000 Hz in most receivers, is extended well beyond those parameters to give you sound with greater clarity and definition. Reviewing built-in features in the 230A, the old heat problem is gone forever with solid-state design, and a special circuit automatically switches the FM tuner to stereo the moment a stereo broadcast is received, then returns to standard FM when the stereo broadcast ends. High speaker damping controls excessive speaker excursions to avoid distortion during sharp bursts of sound and clean, tight bass results. Simple front panel switching eliminates complex and expensive external switching devices, and headphone receptacle is provided for private listening. Precise tuning meter movement clearly indicates strongest and clearest AM and FM response. Dial illumination shows AM or FM dial scale independently for best visibility and entire top section of receiver turns black when switched to phono mode but on-off switch pushbutton glows red while power is on. Specifications are FM sensitivity IHF 2.7 uv; stereo separation 30 db; FM

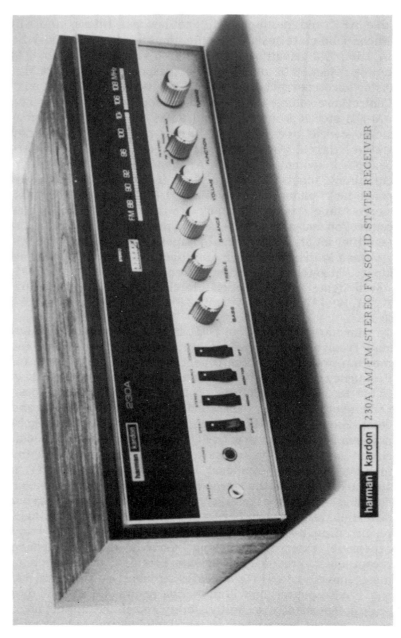

Fig. 2-7. Harman-Kardon model 230A AM-FM-stereo FM solid-state receiver. (Courtesy Harman-Kardon.)

harmonic distortion 0.6 percent; image rejection better than 40 db; damping factor 25:1. A pictorial view is in Fig. 2-7.

Harman-Kardon model 630, AM-FM stereo FM solid-state receiver is twin powered for closely regulated power supply voltages which ensure cleaner and more transparent sound. This is a high performance stereo receiver in the medium-price field, featuring special filtering of SCA, newly designed FM muting, and illuminated on-off pushbutton switch. Special front panel switch and rear panel jacks permit the owner to connect Dolby B processor for use with Dolby B FM broadcasts. The front panel switch changes the de-emphasis network to conform precisely with Dolby B broadcast requirements. The large illuminated tuning meter facilitates tuning exact center of station for best quality and lowest noise. Features include flywheel tuning action, stereo headphone receptacle, and dual concentric tone controls for bass and treble. Heavy duty, ultra-wideband silicon output transistors are operated at 50 percent capacity for cool performance, and DC coupled audio extends lower amplifier response to almost zero Hz with practically no phase shift. Tape monitor facility permits listening to tape being recorded, and two tape recorder outputs on rear panel permit user to connect reel to reel to tape deck and cassette deck simultaneously. Specifications include sensitivity FM 1.9 microvolts (uv) IHF; S-N ratio 70 db; capture ratio 2.5 db; image rejection 50 db; distortion 0.6 percent mono, 0.7 percent stereo; power output 30-30 watts RMS both channels driven into 8 ohms from 20 to 20,000 Hz at less than 0.5 percent THD (120V AC, 50-60 Hz); peak power in excess of 60 watts per channel, both driven; damping factor 40:1; frequency response below 4 Hz to beyond 70,000 Hz plus or minus 0.5 db at normal power. Dimensions are 4¾" H, 12" W, 13¾" D. Weight is 28 pounds. Front panel controls are illuminated pushbutton on-off switch; stereo headphone receptacle; two room stereo speaker switching; tape monitor switch; FM muting switch; high cut filter; stereo to mono switch; contour switch; Dolby B switch; dual concentric bass controls for L and R channels; dual concentric treble controls for L and R channels; zero to infinity balance control; close tracking volume control; function selector switch; large tuning knob; illuminated function call outs; stereo indicator light; large AM-FM tuning meter. See Fig. 2-8 for pictorial view of the 630 model receiver.

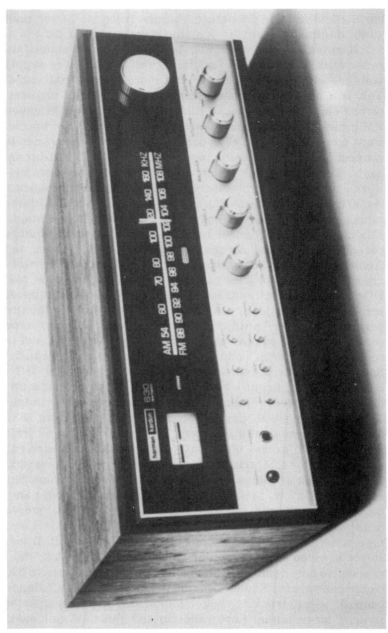

Fig. 2-8. Harman-Kardon model 630 AM-FM-stereo FM solid-state receiver. (Courtesy Harman-Kardon.)

Harman-Kardon model 930, AM-FM-stereo FM receiver is conservatively rated at 45-45 watts RMS with both channels driven into 8 ohms. Two separate power supplies for precise regulation under widely varying conditions enable this instrument to deliver ample power for the low, heavy bass passages. The two power supplies offer complete isolation of both amplifier sections with each channel having its own power transformer and electrolytic filters. Special features include low noise junction-gate field effect transistors in the FM front end for optimum sensitivity; four-gang tuning capacitor for better selectivity; FM muting circuit; special SCA filter for clean stereo signal; illuminated pushbutton on-off switch; quick-action signal strength meter; antenna terminals for 300 or 75 ohm input; monopole AM antenna; flywheel tuning; stereo headphone receptacle; built-in headphone protection against overload; six-pole ceramic IF filter; IC noise and distortion limiting; silicon output transistors operated at 50 percent capacity; DC coupling eliminates phase shift and extends low frequency response; specially designed multiplex circuit yields exceptional stereo separation; widest frequency response; rugged construction; facility to monitor two recorders; provision for two phono systems (low Z); illuminated dial and pointer with blackout in phono or aux. modes; preamplifier and power amplifier may be separated. Rear panel connections provide two low level phono inputs; knurled grounding screw; two sets of high level inputs; two tape recorder outputs; two sets of tape recorder monitor inputs-aux.; AM and FM antenna inputs 300-75; connections for two sets of stereo speakers; patch cords across preamplifier outputs and amplifier inputs (may be removed to use preamplifiers with external power amplifiers or quadraphonic processors); two AC receptacles (one switched); AC fuse and two speaker fuses. Specifications are FM sensitivity IHF 1.8 microvolts; selectivity 50 db; S-N ratio 70 db; image rejection 90 db; stereo separation 38 db; audio response 15 to 30,000 Hz; total harmonic distortion, mono 0.5 percent, stereo 0.6 percent; amplifier bandwidth below 10 to beyond 40 kHz at less than 0.5 percent THD into 8 ohms; amplifier frequency response below 4 to 70 kHz plus or minus 0.5 db at normal power; hum and noise 85 db below rated output. Dimensions are 4¾" H, 17" W, 13¾" D. Weight is 29 lbs. See Fig. 2-9 for pictorial view of the model 930 receiver.

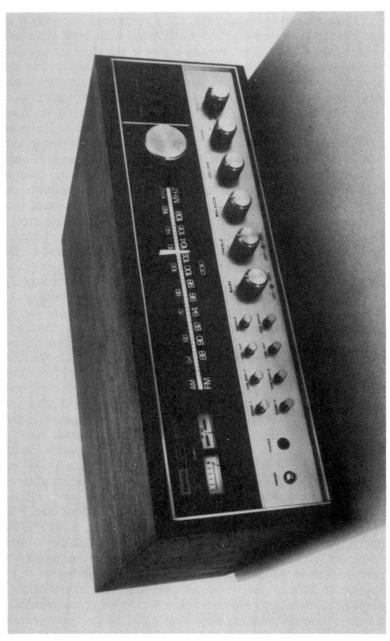

Fig. 2-9. Harman-Kardon model 930 twin powered AM-FM-stereo FM solid-state receiver. (Courtesy Harman-Kardon.)

AR FM is an integrated FM tuner, preamplifier-control and power amplifier, of all silicon solid-state design. Applicable IHF test standards are used for the following specifications: FM sensitivity 2.0 microvolts (hush control off); S-N ratio 65 db; total harmonic distortion less than 0.5 percent mono or stereo; frequency response 20 to 15,000 Hz; selectivity 55 db or more; separation 35 db. The FM tuner section uses FET front end with 4-section capacitor for tuning. Other features include multisection IF filter, integrated circuit IF amplifiers, interstation noise suppression, low distortion shunt multiplex detector, and automatic mono-stereo FM switching. The control amplifier section has a power output for each channel (both driven) of 60 watts RMS into 4 ohms, 50 watts RMS into 8 ohms, or 30 watts RMS into 16 ohms. Harmonic distortion is less than 0.5 percent and frequency response plus or minus 1 db, 20 to 20,000 Hz. Features in the amplifier section include DC driver clamping circuit providing very clean clipping and recovery from overloads which increases the usable dynamic range. The built-in null circuit provides quick, accurate means for channel balancing; and stable, trouble-free operations with any conventional make or design speakers is ensured without damage to transistors from overload. The receiver is pictured (front view) in Fig. 2-10.

Kenwood model KR-7070, FM-AM stereo receiver features automatic-remote tuning with a touch of the tuning bar at either end for left to right scan or right to left scan. Stations are selected automatically along the way on either the main tuning bar or the remote tuning bar. Flywheel manual tuning is provided with a flick of the front panel switch, and a large illuminated tuning meter permits quick pinpoint tuning. Other features are 4 ICs and crystal filter FM IF amplifiers for exceptional selectivity and 1.5 db capture ratio; 3 FETs with 4-gang tuning to provide superior FM front end sensitivity; and 300 watts (IHF) power output to drive any low efficiency speakers. Specifications for FM tuning section are IHF sensitivity 1.5 microvolts (uv); frequency response 20 to 15,000 Hz; harmonic distortion, mono 0.4 percent, stereo 0.6 percent; S-N ratio 70 db; selectivity 75 db; image rejection better than 100 db; stereo separation 35 db. Amplifier section specifications are dynamic power output IHF 300 watts at 4 ohms or 220 watts at 8 ohms; continuous power with both

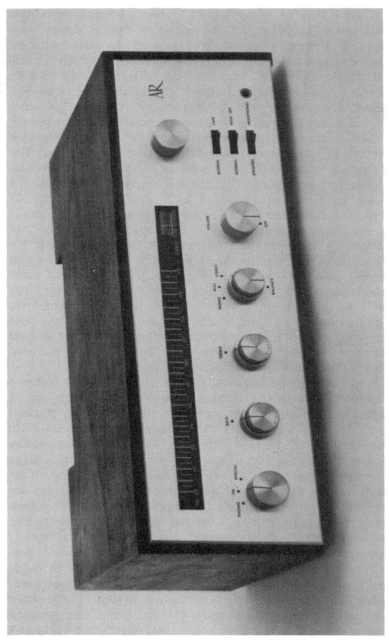

Fig. 2-10. AR FM receiver. (Courtesy Acoustic Research, Inc.)

channels driven, 95-95 watts at 4 ohms, 80-80 watts at 8 ohms; power bandwidth IHF 10 to 30,000 Hz. Rear view layout of the receiver is shown in Fig. 2-11. Size is 6½" H, 17" W, 15" D.

Marantz model 120, AM-FM stereo tuner has a built-in oscilloscope to assure perfect tuning for optimum reception. This is much easier read than three separate meters indicating signal strength, center of channel, and proper antenna orientation to minimize multipath distortion. Gyrotouch tuning, illuminated dial pointer, and pushbutton function switches are among the numerous features. Specifications are FM sensitivity 1.4 uv; selectivity 80 db; capture ratio 1.5 db; frequency response plus or minus 1 db, 20 to 15,000 Hz; image rejection 90 db; stereo separation 42 db; harmonic distortion, mono 0.15 percent, stereo 0.25 percent; dimensions 5¾" H, 15⅜" W, 13" D. Weight 22 lbs.

Fisher model 390, AM-FM stereo receiver represents the lowest cost receiver with electronic tuning in the Fisher line. Tune-O-Matic Memory Tuning instantly selects any of five

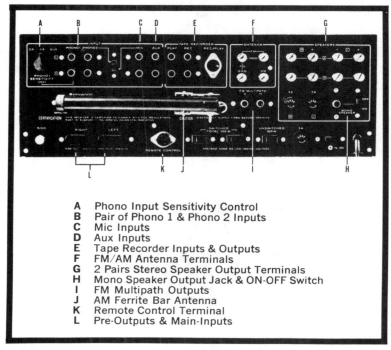

A Phono Input Sensitivity Control
B Pair of Phono 1 & Phono 2 Inputs
C Mic Inputs
D Aux Inputs
E Tape Recorder Inputs & Outputs
F FM/AM Antenna Terminals
G 2 Pairs Stereo Speaker Output Terminals
H Mono Speaker Output Jack & ON-OFF Switch
I FM Multipath Outputs
J AM Ferrite Bar Antenna
K Remote Control Terminal
L Pre-Outputs & Main-Inputs

Fig. 2-11. Rear view layout (Kenwood model KR-7070).

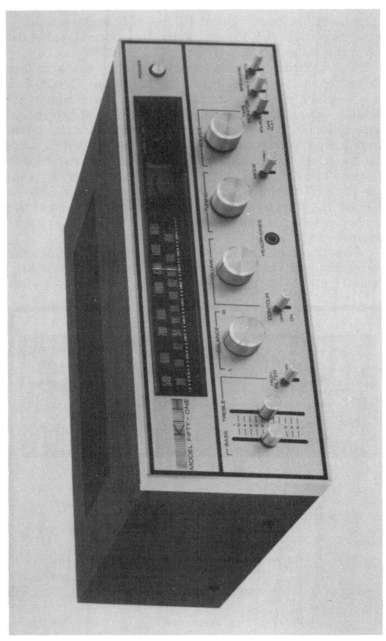

Fig. 2-12. KLH model 51 AM-FM-FM stereo receiver. (Courtesy KLH Research & Development Corp.)

favorite FM stations and also features AFC switch, FM local attenuator switch, and muting switch. Specifications include FM sensitivity IHF 2.0 microvolts (uv); selectivity 45 db; S-N ratio 65 db; image rejection 55 db; stereo separation 38 db; music power IHF plus or minus 1 db 140 watts; RMS power 45-45 watts at 4 ohms; harmonic distortion 0.5 percent; frequency response 20 to 20,000 Hz; power bandwidth 20 to 25,000 Hz. Dimensions are 4⅝" H, 15½" W, 12¾" D.

KLH model 51 stereo receiver is designed as a complete control center aimed at exploiting the capabilities of the many modern, low-cost, high-quality acoustic suspension speakers such as the KLH Model Seventeen, KLH Model Thirty-Two, and KLH Model Thirty-Eight. Features are all silicon solid-state, FM FET front end, and IC limiter along with many others. FM tuner section has a usable sensitivity of 2.5 uv; S-N ratio 65 db; image rejection 55 db; selectivity 50 db; stereo separation 35 db; amplifier RMS with both channels operating 20-20 watts into 8 ohms; IHF music power into 8 ohms 25-25 watts with both channels operating; harmonic distortion THD 0.5 percent; power bandwidth at 8 ohms 15 to 30,000 Hz; frequency response 10 to 35,000 Hz plus or minus 2 db; damping factor at 8 ohms greater than 20. Dimensions are 6¼"H, 17"W, 15⅜"D. Weight is 21.5 lbs. See Fig. 2-12.

Altec model 725A, AM-FM stereo receiver is designed for optimum quality and flexibility with the exclusive Varitronik Tuner. Other features are advanced muting circuit, 4 FETs and balanced Varicap tuning, and combination Butterworth and crystal filters for greater stereo separation, capture ratio, and selectivity, plug-in circuits, and fail-safe protection system with black-out dial plus indicator lights on all functions, separate tuning meters for signal strength and center tuning; stereo tape recorder inputs and outputs on front panel; accessory jacks for the Altec Acousta-Voicette Stereo Equalizer; spring loaded terminals for main and remote speakers; indicator lights and pushbutton controls on all functions. Specifications for the FM tuner section are IHF sensitivity 1.8 uv; capture ratio 1.3 db; stereo separation better than 40 db; harmonic distortion 0.3 percent; frequency response plus or minus 1 db 20 to 15,000 Hz; image rejection 90 db; amplifier power output RMS both channels driven 8 ohms 60-60 watts; distortion THD less than 0.3 percent; frequency response 20 to 20,000 plus or minus ½ db; power bandwidth 15 to 25,000 Hz; damping factor 25; dimensions, 5" H, 17¾" W, 16½" D.

Sansui model 8 AM-FM stereo tuner-amplifier represents a totally new concept in the manufacture of quality stereo receivers. Features include a high sensitivity front end using three expensive low-noise MOSFETs, protector diodes, and a 4-gang frequency linear variable capacitor. This ensures a dramatic edge over conventional receivers in intermodulation distortion and image ratio characteristics. The IHF FM sensitivity of 1.7 microvolts permits weak stations to come in with full, hi-fi clarity. Other notable points are the three-IC IF amplifier; sharp-cutting multiplex carrier leak filter; FM muting switch; adjustor; FM linear scale; large tuning meters; 300 and 75 ohm antenna terminals; FET AM tuner; unique pantograph antenna; smooth tuning dial pointer; 3-stage equalizer amplifier; two tape deck accommodation; negative feedback control amplifier; triple tone control circuit; negative feedback high and low filters; direct coupled power amplifier; two 8,000 uf power supply capacitors; jumbo heat sink; complete protection circuit; independent preamp and power amplifiers; handles three sets of speaker systems; stereo balance circuit; separate power circuits; separate block chassis construction; and full system accessory circuit. The 2-stage LC filter with sharp cutoff characteristics for the carrier leak filter in the multiplex circuit prevents the multiplex pilot from affecting the audio signals to cause beat interference and unpleasant noise or intermodulation distortion. It also prevents beat interference with the bias frequency of a tape recorder if the multiplex broadcast were to be recorded. But the sharp cutoff LC filter, designed to combat carrier leakage, results in drastically reduced beat interference and intermodulation distortion. The safety of the direct coupled amplifier depends upon the level of temperature compensation and the performance of the differential amplifier that protects the power transistors. This model employs not only extra transistors for temperature compensation, but incorporates a special differential amplifier stabilizing circuit to stabilize its two-stage differential amplifier. In addition, six quick-acting fuses and a special power limiter circuit give the power transistors further protection in the unlikely event of an overcurrent. Consideration has also been given to the superior capability of the

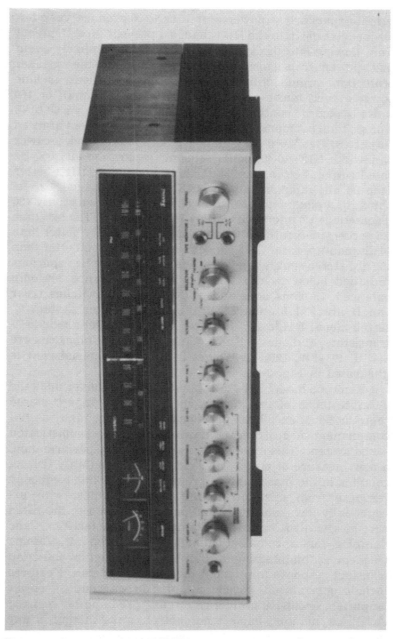

Fig. 2-13. Sansui model 8 AM-FM stereo receiver. (Courtesy Sansui Electronics Corp.)

output-capacitorless power amplifier to reproduce very low frequencies down to the DC range. To prevent direct current from flowing into the speaker systems in the unlikely event that power transistors should be damaged, the speaker protection circuit utilizes an SCR. Specifications include power output, music power IHF 200 watts (4 ohms), or 160 watts (8 ohms); continuous power, 80-80 (4 ohms), or 60-60 (8 ohms); total harmonic distortion less than 0.4 percent at rated output (60 Hz; 7,000 Hz equals 4:1 SMPTE method); power bandwidth IHF, 10-40,000 Hz; channel separation at 1,000 Hz rated output, better than 60 db; hum and noise (IHF) better than 90 db; input impedance 50K; load impedance 4 to 16 ohms; damping factor, 60 with 8 ohm load; FM sensitivity (20 db quieting) 1.4 microvolts, IHF 1.7 microvolts; THD less than 0.5 percent; S-N ratio better than 65 db; selectivity better than 60 db; capture ratio (IHF) 1.5 db; image rejection better than 100 db; stereo separation better than 35 db at 400 Hz; spurious radiation below 34 db; antenna input impedance 300 ohm balanced, 75 ohms unbalanced. Controls and switches (rear panel) are FM muting level; mode of speaker system C, stereo-mono; FM level; AM level. Semiconductors include 77 transistors, 41 diodes, 5 FETs, 3 ICs, 4 Zener. Dimensions are 5½"H, 17½"W, 12⅞"D. Weight is 37.4 lbs. Pictorial view is illustrated in Fig. 2-13.

Kenwood model KR-7200 FM-AM stereo receiver adds the ultimate touch of luxury to your home with superb stereo reproduction, excellent broadcast reception, and a full complement of controls for an expansive and sophisticated sound system. This model offers extraordinary performance, a new measure of power, and exceptional reliability. Direct coupling in the power amplifier stage ensures flat frequency response from ultra-low to the highest ranges, with exceptionally low distortion at all power levels up to the rated output. It also improves damping characteristics, and damping factor remains uniformly high even at subsonic frequencies, maintaining faithful bass reproduction and crisp transient response throughout the audio spectrum. Features include guaranteed power output; direct coupled power amplifier; exclusive protection circuit; preamplifier section; dual tape monitor; triple tone control; mike mixing in any mode; separate use of preamp; one-touch speaker terminals; linear FM scale; newly-developed DSD circuit; special FM IF

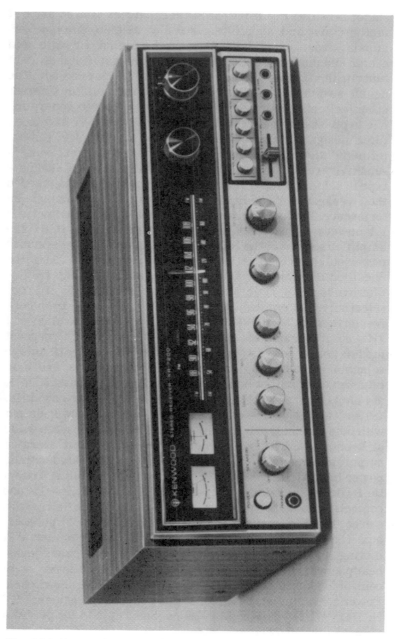

Fig. 2-14. Kenwood model KR-7200 FM-AM stereo receiver. (Courtesy Kenwood.)

stage; sensitive FM front end. Other features offered are FM muting; low and high filter; 300 and 75 ohm antenna terminals; front panel headphone jack; signal strength and center channel tuning meters; input indicator lights; AM tuner with separate RF and IF sections; walnut cabinet. The double-switching demodulator (DSD) in the multiplex circuit effectively eliminates the deterioration of stereo separation due to phase drift between main and sub-signals. A new block filter design efficiently reduces beat interference and intermodulation distortion due to carrier leakage. Front end FM sensitivity results from the use of 3 FETs including a dual gate type to achieve a superb 1.6 microvolt rating and bring in even the weakest FM stations without adjacent channel interference. While many amplifiers are unable to sustain their full power measured at 1 kHz in the range below 50-60 Hz, the amplifier section of the KR-7200 fully guarantees a continuous power output of 55 + 55 watts at 8 ohms from 20 to 20,000 Hz. Specifications for the FM tuner section are sensitivity IHF 1.6 microvolts; S-N ratio 68 db; capture ratio (IHF) 1.5 db; selectivity, alternate channel (IHF) 75 db; frequency response 20-15,000 Hz +0.5 to —1.5 db; stereo separation at 1 kHz 40 db; harmonic distortion, 400 Hz at 100 percent modulation mono 0.4 percent, stereo 0.6 percent; image rejection 90 db. Amplifier section specifications are continuous power output both channels driven, 110 watts, 55 + 55 at 8 ohms from 20-20 kHz; 120 watts, 60 + 60 at 8 ohms at 1 kHz; 150 watts, 75 + 75 at 4 ohms at 1 kHz; dynamic power output with both channels driven (IHF) 260 watts at 4 ohms, 210 watts at 8 ohms; total harmonic distortion 0.5 percent at rated output; intermodulation distortion 0.5 percent at rated output; power bandwidth (IHF) 10-30 kHz; damping factor at 8 ohms 50. Switches include speakers off, A, B, C, A+B, A+C; tone controls 2 db step-type, bass, mid, treble; mode left, right, stereo, reverse, mono; selector AM, FM, phono 1, phono 2, aux. 1, aux. 2; pushbuttons for power, A-B tape monitor, FM muting, loudness, low filter, high filter; front panel jacks-headphones, rec-play tape B, microphone; slide lever, level control for microphone; AC outlets, 2 switched, 1 unswitched. Power consumption, 350 watts at full power. Dimensions: 5¾" H, 17⅛" W, 14" D. Weight is 29 lbs. See Fig. 2-14 for pictorial view of the KR-7200.

The Pioneer model SX-727 AM-FM stereo receiver features increased performance with greater power, un-

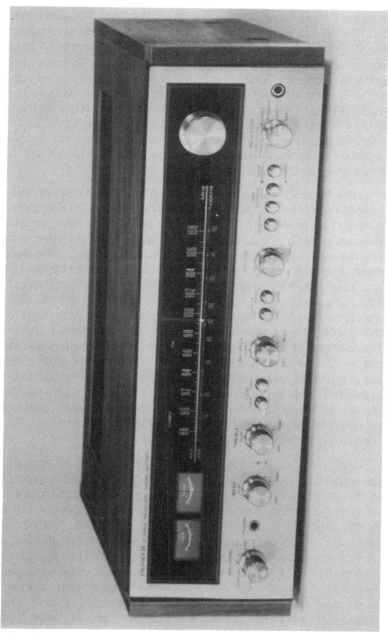

Fig. 2-15. Pioneer model SX-727 AM-FM stereo receiver. (Courtesy U.S. Pioneer Electronics Corp.)

surpassed precision and many new features for total versatility. Power output 195 watts IHF employing direct coupled amplifiers and dual power supplies. It offers improved bass while transient, damping, and frequency responses are greatly enhanced. New circuit protects your speakers against possible damage and DC leakage, and the ultra wide linear dial scale facilitates tuning to the nth degree. High and low filters, FM muting, click-stop tone controls, mode lights, audio muting, and full range of connections for every conceivable accessory are included. The price is $349.95. The SX-727 is illustrated in Fig. 2-15.

The Pioneer model SX-2500 AM-FM stereo receiver with Scan-o-matic tuning features a motorized signal-searching scan device in either AM or FM. A small remote unit permits the user to control volume level and tuning from the other side of the room as well as at the receiver. Arrow-light indicators show direction of scan and point of selection. Selector switch provides AM, FM mono, FM auto, phono, aux. 1, and aux. 2. The FM auto switches to stereo automatically when available. Power output is 340 watts with polarized jacks and plugs to avoid any chance of incorrect phasing or accidental shorts. Outputs are protected by special circuits and fuses. Separate outputs for preamplifier and inputs for power amplifier are provided along with a slide switch to connect for conventional operation. This set offers exceptional FM sensitivity with the ability to bring in weak stations without adjacent channel interference. It is priced at $549.95, including remote unit. The SX-2500 is pictured in Fig. 2-16.

The Altec model 710A, AM-FM stereo receiver recently announced offers the many features and performance typical of the quality found in model 714A. Integrated circuits and all-silicon transistors provide the exceptional specifications in a medium-priced receiver. Power RMS is 30 watts with both channels driven at 8 ohms, and less than 0.5 percent distortion across the entire bandwidth of 30 to 20,000 Hz. Excellent sensitivity on FM of 2.5 microvolts; pushbutton function switches, stepped type tone controls; interstation muting; main and remote speaker switch; headphone jacks; and signal strength meter are but a few of the many features. The price is $349.95. The pictorial view of this latest addition to the Altec line is shown in Fig. 2-17.

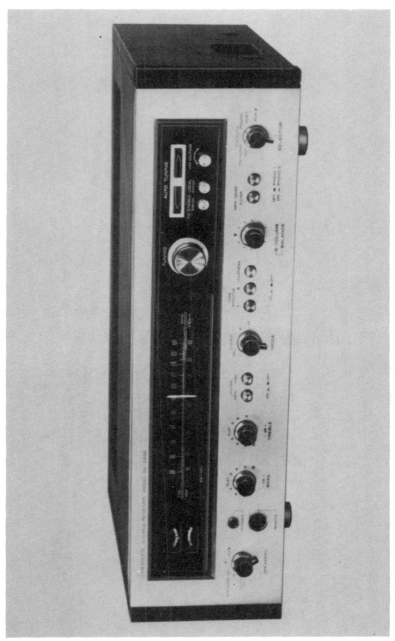

Fig. 2-16. Pioneer model SX-2500 AM-FM stereo receiver with Scan-o-matic tuning. (Courtesy U.S. Pioneer Electronics Corp.)

Fig. 2-17. Altec model 710A AM-FM stereo receiver. (Courtesy Altec Lansing.)

Power Requirements

You should have sufficient power to fill the listening area with sound at a reasonable loudness, and personal taste will vary widely on what is reasonable. First a word of caution against damaging your speaker system, you can't feed a 10 watt speaker with a 100 watt receiver-amplifier and play safe by keeping the volume level down. There are simply too many things that can happen to upset your calculations, so by all means, play it safe. The variables in selecting the power for your amplifier needs are the speaker system, the listening area, your personal level needs and in some cases what the neighbors will tolerate. Discuss the matter with your hi-fi dealer, and make sure the speaker system is matched to the choice you eventually make. Chapter 7 could help in evaluating the latter.

Cost Considerations

There should be some ratio for any stereo system expenditures, but by any formula the AM-FM receiver should get at least half of the total. Check the specials, close-outs, and other types of price reductions but do not cut your stereo budget short in this all-important "control center" area.

Chapter 3

Four Channel Sound

Four channel sound is a modern advancement in sound technology and adds considerably to the realistic as well as natural characteristics of the reproduced music. It enables you to get into the center of reproduction by surrounding yourself with the rendition right in your own home, even as though you were there—front row center. This quadraphonic sound creates the concert hall depth to reproduce music in your living room with uncanny acoustical illusions.

Four channel stereo is available on records, tapes, or over the air in regular FM broadcasting, but there are several ways of hearing it. These are divided into three basic methods although approaches vary with different manufacturers. The true method is labeled **discrete** four channel stereo as it actually provides four individual sources of sound (recorded and played back as such), in addition to the 4-channel source, you must have four amplifier channels with speakers for each. However, if you currently have regular stereo, you may easily convert to this system by adding an amplifier and two speakers for about $125 or so. The **Matrix** four channel stereo method provides recorded music on four channels but blended into two to permit playback on conventional stereo equipment with the addition of SQ decoder plus 2 channels of amplification with their individual speakers. The final or third system is the **derived** four channel stereo which utilizes conventional 2-channel material but by electronic "manipulation" actually simulates most of the four channel effect. So by adding the 4-channel adapter and two speakers to your conventional 2 channel stereo, the "simulated" effects of 4-channel sound may be enjoyed for as little as $60. As you may anticipate, there is a wide difference of opinion as to the best of the three systems, and you may have to reserve that decision for your own personal taste. There is a difference in the cost involved, but it is possible to select your own choice of modes in some of the more sophisticated units.

As the stereo speakers, left and right, permitted us to enjoy the full orchestra instead of a compromise from the center position, the four channel sound includes the rear left and rear right positions to give the far more satisfying feeling of true depth...another dimension in sound as reflected from the rear and side walls. The rear channels with their slight delay make the circle of sound much more complete so that you are definitely surrounded by the sound waves in their full richness.

UPDATING YOUR STEREO WITH A CONVERTER

The Electro-Voice EVX-4 Decoder is a compatible system for four channel stereo and may be easily connected to your present stereo amplifier or receiver as shown in Fig. 3-1. A pictorial view of the decoder is also shown in Fig. 3-2. The system makes it possible to broadcast and record four-channel stereo on existing two channel equipment by the use of a simple encoder at the broadcast or recording source, and simply adding the inexpensive decoder in the home. Needless to say, the encoded signal is completely compatible with equipment already in use and may be received and reproduced as two-channel sound without the decoder, just as color television programs are received in black and white on monochrome receivers.

The addition of the decoder to an ordinary 2-channel stereo system does require four amplifier channels with speakers, but no change in the stereo tuner, record, or tape player. The EVX-4 expands the existing source to four separate channels of sound with no two alike. Even without encoding, two channel material will demonstrate an advantageous quality. The unit actually provides four different from two without splitting frequency, tubing delay-lines, or any form of mechanical magic. This is possible as such material is normally available on original recordings even though mixed down to a pair of channels, and the decoder has the faculty of recreating that original information.

In order for the four independent channels to be reproduced, two stereo amplifiers with each having a pair of channels, and four speakers are needed. Using the preamp and input selector sections of the conventional stereo to feed the decoder, while the power amplifier sections are used to

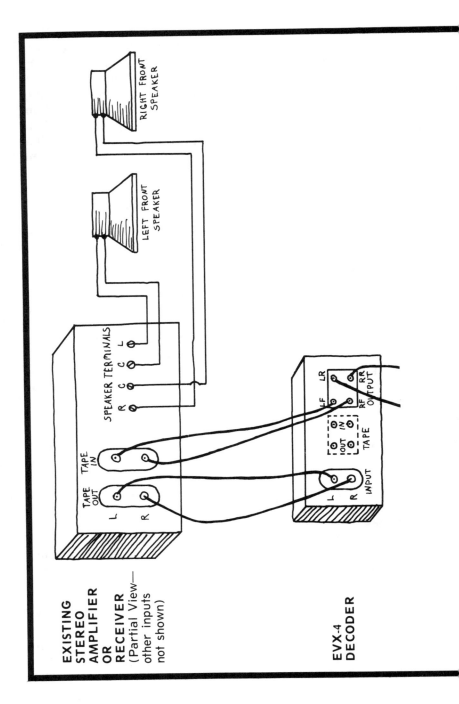

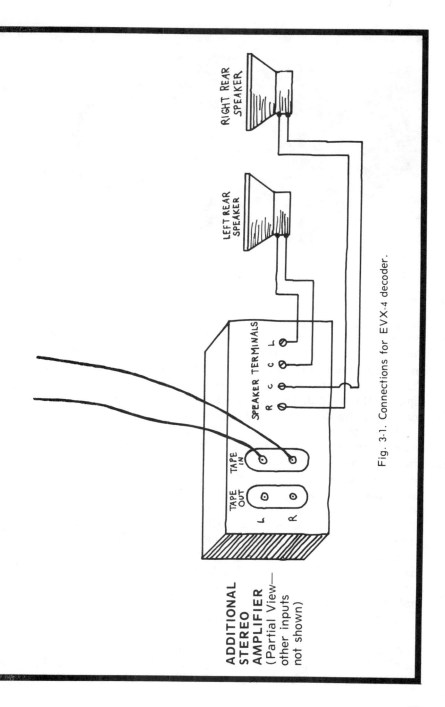

Fig. 3-1. Connections for EVX-4 decoder.

Fig. 3-2. Electro-voice EVX-4 decoder. (Courtesy Electro-Voice, Inc.)

energize the left front and right front speakers, the tape monitor or source switch interrupts the usual signal path through the amplifier to offer a new signal from outside the unit, independent of the input selector switch, as clarified in Fig. 3-3. If a separate control permits the selection of tape input without changing the normal input switch, the connection of the decoder is a simple matter.

Insert the EVX-4 in your system as desired, ahead of the power amplifier such as between the preamp and the power amplifier, or by using the tape monitor connection. It should be noted that good performance in the 4-channel system is far less critical than in the two-channel stereo even though best performance is assured with all four amplifier channels and all four speakers the same. Satisfactory results may still be expected with lower power in the rear amplifier and less frequency range in the rear speakers. The explanation being that the rear channels are aided by the front channels in the low frequency area. However, it is important to have closely matched speakers for the front pair, and for the rear pair.

Decoder output can be no greater than the input voltage from the tape jacks of the front amplifier with the decoder gain control fully advanced. Thus the input sensitivity of the rear amplifier must be high enough that the decoder output will afford satisfactory drive, and if sufficient when connected directly to tape output jack of front amplifier, the decoder will work with it properly.

SYSTEM WIRING

Wiring instructions will be simplified by referring to the unit that operates the front pair of speakers as the **front amplifier**, and the section feeding the rear speakers as the **rear amplifier**. The various combinations possible such as tuner with amplifier, integrated amplifiers, preamps, power amplifiers, compact systems, and consoles that may comprise your system will be easier to adapt to the four channel system with these terms.

The program source, i.e., phono, tuner, tape, etc. will connect to the front amplifier and a more complete unit as a front unit. Where two similar units are used, the one having the better capability from the standpoint of output or lower distortion should be the front unit. By the same token, if two pairs of speakers are available with one better than the other, the better pair should be used as front speakers. Hookup directions must be followed carefully and do take your time to avoid confusion. There are several connections to be made and the step by step procedure will eliminate the risk of error. Make all the L (left) channel connections first and then do everything over again the same way as before for the R (right) channel. You may even mark your front and rear amplifiers with masking tape stuck on the back for ready reference while wiring. Never hesitate to label units this way, even the technician makes notes when repairing or replacing parts in these units to avoid making a mistake.

Connection Procedure

Use convenient lengths of shielded audio cable, connect front amplifier's tape out jacks to the decoder input jacks.

Connect decoder front output jacks to tape in jacks on the front amplifier, and now you should have cables going to and from the front amplifier, and connected to the decoder input and front output jacks.

Connect decoder rear output jacks to rear amplifier input, and use tape in jacks if available. Otherwise, use auxiliary or other high-level inputs.

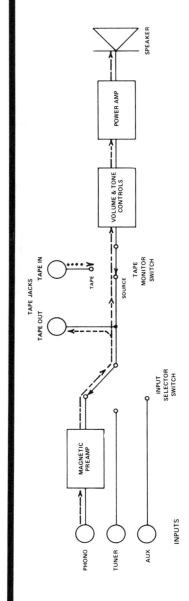

a. With the Tape Monitor switch in the Source position, the source selected by the Input Selector switch plays through the speaker. The source signal also appears at the Tape Out jack.

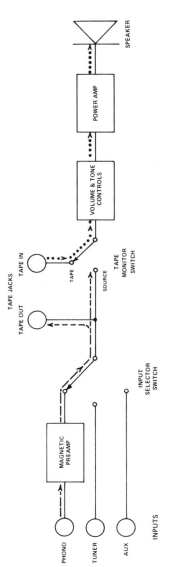

b. With the Tape Monitor switch in the Tape position, the Tape In signal plays through the speaker. The signal at the Tape Out jack is still the source signal.

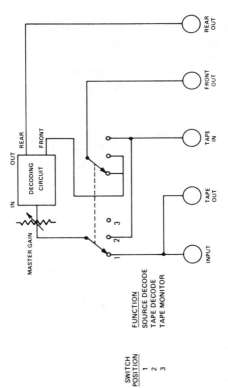

Fig. 3-3. Connections to tape jacks.

Connect the four speakers to the proper amplifier output terminals, just like an auto seat layout: driver left front, passenger right front, passenger behind driver—left rear, and passenger behind passenger—right rear. The left rear speaker is connected to the left speaker terminals of the rear amplifier, and the right rear speaker to the right speaker terminals of the rear amplifier. All speakers must be connected in phase as shown in the instructions.

If using a tape recorder in your system, connect it to the tape in and tape out jacks on the rear of the decoder. These jacks duplicate the jacks previously used on the front amplifier and permit decoding of the output of your tape recorder into four channels.

Connect AC line cord from decoder to convenient outlet or if a switched outlet is handy on one of your amplifiers, power may be applied through the amplifier and the decoder's master gain control left turned up at all times. When switched AC outlets are available on both front and rear amplifiers, plugging the rear amplifier into the front amplifier and decoder into rear amplifier, permits entire system to be turned on with front amplifier power switch.

Operation

After decoder is connected, amplifiers' tape monitor switches are placed in tape position, or aux. position on rear amplifier, is used. Output of decoder is connected into all four amplifier channels. With amplifier volume control at minimum, decoder master gain control is fully advanced (clockwise), and decoder function switches are placed in Source Decode position. Advancing amplifier volume controls (clockwise) will provide the desired sound level in the room and permit adjustment of the front-to-back balance. Setting the volume controls for the loudest sound desired, with front and rear sound levels about the same, the master gain on the decoder will lower the sound from all four speakers equally when turned counterclockwise. After the preliminary adjustment, rear amplifier volume control will allow adjustment of front-to-rear balance by adjusting the rear sound level up or down slightly to move the balance point frontward or backward in the room for best subjective effect. Enhancement of a two-channel classic should endeavor to recreate the ambient sound in the hall. This would often be possible by reducing the rear amplifier level somewhat with less treble.

Left-right balance controls operate normally, but with less need to adjust the balance in either front or rear. To return to normal straight-through channel operation of the front speakers, return front amplifier tape monitor switch to the source position and on the rear amplifier, either switch back to the source, or switch out of aux. according to your individual hookup; or simply turn down the volume control (counterclockwise).

If your tape recorder is connected to the decoder, conventional two-channel tapes may be recorded and played through either the two-channel or four channel system. Signals to the tape recorder are exactly the same as those from the tape out jacks on the front receiver where the recorder is usually connected. As the decoder switch is in the tape monitor position, the tape recorder plays back through the front amplifier in conventional two-channel mode. Moving the function switch to the tape decode position causes the output of the two-channel tape recorder to be decoded into four-channels exactly as an FM broadcast is decoded in source decode position. Naturally when using the discrete four-channel tape unit as part of your system, the decoder is not necessary for four-channel playback, and the four outputs are connected to auxiliary inputs of the amplifiers.

More Shortcuts to Quadraphonic Sound

There is little doubt that change comes hard with most of us, but when we are finally sold on a new technique or system, we stand up for it quite strongly. Listen to 4-channel, quadraphonic sound, not once or twice but every chance you get and make them sell you on this new technique. Actually you can do this at home in some of the inexpensive ways that follow.

The advantages of matrixing need only to be reviewed to emphasize the logic of this mode of personally evaluating the possibilities of four channel sound. The method is equally applicable to disc recording, tape recording, and FM broadcasting as well as any other two channel medium. A simple inexpensive decoder works for all media and there is no need to discard or replace phonographs, cartridges, tuners, amplifiers or any other stereo equipment, they are all converted as they currently exist as soon as the decoder is added. This lack of expense is important, at least while you are checking the idea thoroughly. If you are sold on the new sound, that is the time to entertain the thought of more sophisticated equipment and the much greater expense. Most decoders are capable of serving as good synthesizers when subjected to conventional two-channel stereo. This is no small asset, since the long transitional period from two-channel to four-channel stereo permits the use of the listener's older records for worthwhile four-channel effect. This immediate advantage also applies to many two-channel FM broadcasts that will be offered during the interim. It is true that there is more than one matrixing system in use. Three significant systems, in alphabetical order are CBS-Columbia SQ, Electro-Voice Stereo 4, Sansui QS. Major differences occur only at the encoding or studio end. At the decoding end which is the only point of concern to the listener, the three are highly compatible. Whatever decoder is used, it will do a very good job of reproducing the records made by any of these systems named above.

Metrotec model SD4A-Q, Universal Four Channel Decoder-Rear Channel Amplifier decodes all new SQ and EV 4-channel records and FM broadcasts plus synthesizes quad sound from any 2-channel source. The matrix-phase shift decoding circuits have been designed to decode all types of the new 4-channel material, and a flick of the switch will give the exact phase shifts and coefficients to match the program material. The powerful 30-watt rear channel amplifier with all controls on a single panel provide everything required, simply add the two rear speakers. Most major record companies are producing 4-channel discs and this unit when connected to your system with two rear speakers will make it possible for you to derive full pleasure from any or all. Complex information already exists on most stereo records and goes undetected since the usual 2 speaker stereo cannot reproduce multi-dimensional signals. Just think of the amount of sound on your record collection that you have never had the opportunity to hear until now. The amplifiers in the model SD4AQ offer an IHF music power of 30 watts at 8 ohms (15 watts per channel) and an RMS power of 20 watts at 8 ohms (10 watts per channel). Other specifications are harmonic distortion 0.8 percent; power bandwidth 15 to 50,000 Hz;

frequency response 10 to 50,000 Hz plus or minus 2 db; S-N ratio 80 db; damping factor 30 to 1; price $149.95. See Fig. 3-4 showing simple connection to an existing receiver. A picture of the SD4A-Q is displayed in Fig. 3-5.

SSI Quadrasizer I is a unique solid-state encoder that permits you to produce quadraphonic recordings on your present 2-channel tape machine. It accepts four microphone or musical instrument inputs. The signal from each microphone is recorded on any 2-channel tape machine by feeding the input from one microphone discretely on the left channel, the input from a second microphone discretely on the right channel, and the input from a third mike recorded on both left and right channels at equal levels, perfectly in phase, providing a true center channel. The input from the fourth mike will result in ambient material for rear channel speakers. The block diagram shows connections in Fig. 3-6. It is priced at $59.95.

SSI Quadrasizer IV is a 4-channel playback adapter or decoder to permit you to synthesize quadraphonic sound from any 2-channel source such as phono player, tape player, FM receiver. This unit senses ambient material already on stereophonic tapes or discs and interprets it for reproduction to rear channel speakers. No additional amplifiers are

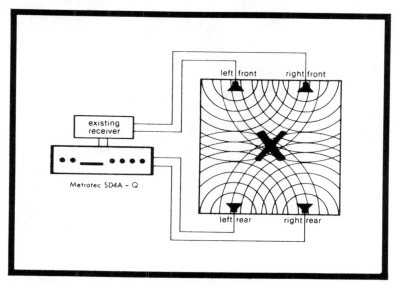

Fig. 3-4. Connecting SD4A-Q.

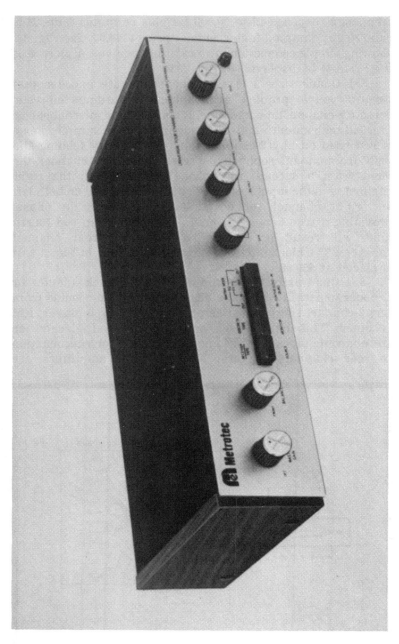

Fig. 3-5. Metrotec 4-channel decoder-rear amplifier. (Courtesy Metrotec Electronics, Inc.)

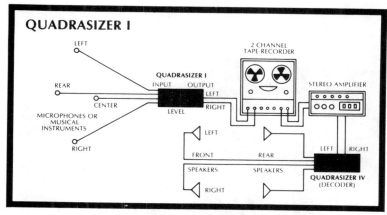

Fig. 3-6. System connection.

necessary but may be used if preferred. It is priced at $29.95, and connections are quite simple as shown in Fig. 3-7.

EV 1224X amplifier (decoder) provides the easy expansion of any current stereo system into a 4-channel system by providing two complete amplifier channels with full controls and the integrated circuit decoder. This unit creates a remarkable enhancement of all conventional two-channel material. In this way, nothing is wasted, the existing stereo system is fully utilized in the conversion, which is quite simple to make by following the simple directions included. The picture view of the Electro-Voice EV 1244X is shown in Fig. 3-8. The decoder is easily bypassed when 4-channel is not

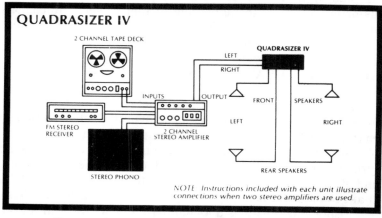

Fig. 3-7. Block connections for (SSI Quadrasizer IV).

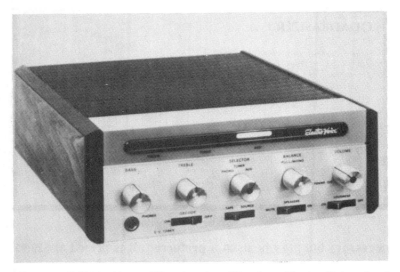

Fig. 3-8. E-V 1244X amplifier-decoder. (Courtesy Electro-Voice, Inc.)

desired as the EV 1244X is equipped with full switching and control capabilities to permit use with existing stereo system for 4-channel tape playback. Either reel to reel (open reel) or cartridge machines may be used or the EV 1244X will function independently as the control center of a fine 2-channel stereo system. The power output of the amplifier is 18 watts per channel and the IHF music power into 4 ohms measures 65 watts or 50 watts into 8 ohms. Frequency response 20 — 20,000 Hz plus or minus 1.5 db at rated power with a harmonic distortion of less than 1.0 percent at that power. Dimensions are 3⅜" H, 8⅜" W, 10¼" D. In order to emphasize switching arrangements, the table of control settings is shown in Fig. 3-9.

ADDITIONAL FEATURES IN OTHER UNITS

A few of the more popular 4-channel receivers and some of their more important features are covered in this section to enable you to consider the advantages which may be applicable to your individual arrangement. Specifications as furnished by the manufacturer are included where available, and it should be remembered that these figures could be changed in later models or even later runs of the same model.

Fisher model 601, is a 4-channel 200 watt AM-FM stereo receiver featuring 2 + 2 decoder to extract hidden ambience

OPERATION	FRONT AMPLIFIER		E-V 1244X Decoder/Amplifier				
	SELECTOR SWITCH	TAPE/SOURCE SWITCH	SELECTOR SWITCH	SOURCE/TAPE SWITCH	VOLUME CONTROL	DECODE SWITCH	TAPE MONITOR
Decode Records	Phono	Tape	AUX	Source	Master	On	Normal
Decode FM Stereo Broadcast	Stereo FM	Tape	AUX	Source	Master	On	Normal
*Decode 2-channel (or encoded) tapes	Any	Tape	Any	Tape	Master	On	Normal
*Play 2-channel tapes without decoding	Any	Tape	Any	Tape	Back only	Off	Tape Mon 2
**Decode 2-channel (or encoded) tapes	AUX	Tape	AUX	Source	Master	On	Normal
**Play 2-channel tapes without decoding	AUX	Source	AUX	Source	Back only	Off	Either
**Play 4-channel discrete tapes	AUX	Source	Any	Tape	Back only	Off	Either
Decode back only (Hall Sound)	Phono or FM Stereo	Source	AUX	Source	Back only	On	Either

*2-channel Tape Machine   **4-channel Tape Machine

Fig. 3-9. Electro-Voice E-V 1244X control settings.

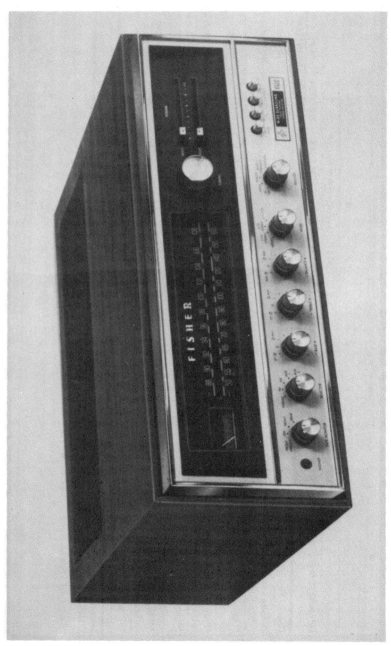

Fig. 3-10. Fisher model 601 4-channel receiver. (Courtesy Fisher Radio.)

information from conventional stereo materials and play it through rear channels to closely approximate discrete 4-channel reproduction. Plug-in multiplex decoder section adaptable to current FM 4-channel broadcasts, muting switch, and high filter are included. The model 601 handles up to eight speaker systems, four main and four remote. Specifications for the FM tuner section are usable sensitivity IHF 1.8 uv; S-N ratio 66 db; selectivity 45 db; image rejection 65 db; harmonic distortion 0.4 percent; stereo separation 35 db; amplifier section: RMS power 36-36-36-36 watts at 8 ohms; music power IHF plus or minus 1 db at 1 kHz 200 watts; harmonic distortion 0.5 percent; frequency response 20 to 25,000 Hz; power bandwidth 25 to 22,000 Hz; damping factor 10. Dimensions are 5¼" H, 17" W, 16½" D. Refer to Fig. 3-10 for pictorial view of the model 601.

Sansui model QR6500, is a 4-channel 280 watt AM-FM stereo receiver with sensitive AM-FM tuner, high power 4-channel amplifier, and a decoder for matrixed 4-channel recordings and broadcasts plus a synthesizer for 2-channel sources. It features continuous power per channel of 37 watts at 8 ohms; power bandwidth 20 to 30,000 Hz; THD (distortion) below 0.5 percent; 1.8 uv sensitivity IHF; exclusive QS phase-shift and phase-modulator circuits for true 4-channel source localization and live sound-field effect; speaker mode switch; selector for 5 pairs of front and rear speakers; master 4-channel volume control; illuminated digital 2 and 4-channel mode indicators; discrete 4-channel input; 4 and 2-channel tape recording outputs and monitors; tape to tape copying; 3 slide controls for total left-right and front-rear balance; stabilized power supply; headphone jacks, front and rear channels; wide linear FM dial; 3-stage ceramic IF filters; FM muting switch and 3 AC outlets.

Lafayette model LR-440, is a 4-channel 170 watt AM-FM stereo receiver with superb AM-FM tuner, 4 amplifiers for discrete 4-channel, and derived 4-channel capability as well. This compatible 4-channel receiver brings full performance to your home with amazing dimensional depth and richness. Discrete 4-channel programs from 4-channel, 8-track cartridges and 4-channel reel to reel tapes plus other sources as they become available. Specifications include FM tuner sensitivity IHF 1.65 uv; stereo separation 40 db; S-N ratio 75 db; harmonic distortion 0.3 percent at 100 percent modulation;

power output 170 watts plus or minus 1 db (42.5 watts per channel) at 4 ohms, 110 watts (27.5 watts per channel) at 8 ohms; frequency response 20 to 20,000 Hz; total harmonic distortion 0.07 percent at 1 watt output, 0.8 percent at rated output; power bandwidth 15 to 25,000 Hz. Dimensions are 4¾" H, 18½" W, 13¾" D. Rear panel has input jacks for 2-channel and 4-channel stereo for tape playback, magnetic and ceramic phono and auxiliary I and II. Main and remote speaker outputs for all four channels with two AC convenience outlets (one switched). Meter indicates signal strength of incoming AM or FM signal.

The paths to 4-channel sound are numerous, and probably the first consideration would be whether or not the stereo equipment currently in use is satisfactory. Assuming that you are happy with your stereo system at this time, the 4-channel decoder-amplifier will let you bring that system right up to date without too much expense. You will also need a pair of matched speakers for the 2 additional (rear) channels, and they do not need to be quite as good as the pair you are now using up front. Some of these 4-channel converters will be described in detail shortly. Now let us look at the opposite condition—you are not satisified with your stereo system as it stands—then by all means go all the way with the 4-channel AM-FM stereo receiver, if the budget can possibly stand it. A few of these complete receivers were described in detail

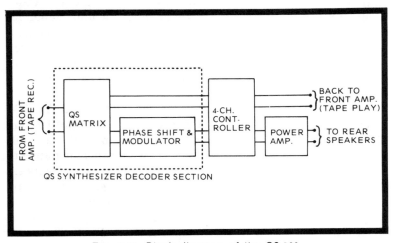

Fig. 3-11. Block diagram of the QS-100.

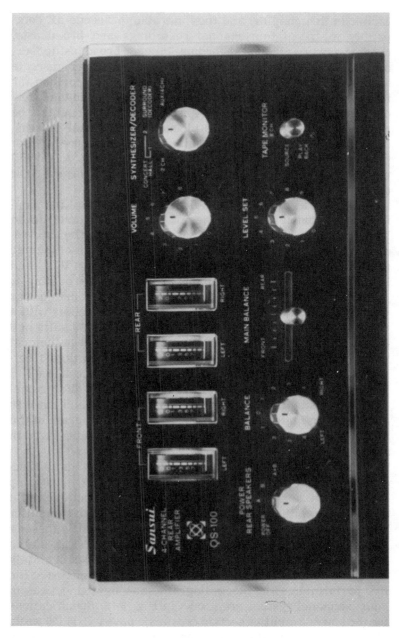

Fig. 3-12. Sansui model QS-100 4-channel converter-synthesizer-amplifier. (Courtesy Sansui Electronics Corp.)

earlier in this chapter, and they are able to decode all compatibly matrixed four-channel recordings and broadcasts, convert any two-channel recording, tape, or broadcast to four, or play any discrete four-channel tape or other discrete source. Most of these 4-channel receivers have provided for future possibilities with extra inputs and facilities to take care of adapters and converters that could subsequently be required. However, if you are blocked for financial reasons, the converter route will provide a good measure of improvement for your inadequate system, and the new sound will please even the more demanding listener to a degree that could be surprising.

The Sansui model QS100, 4-channel rear amplifier converts the 2-channel stereo to the much richer 4-channel stereo by merely adding another set of speaker systems. The entire 4-channel synthesizer-decoder section is an all-electronic printed circuit board using specially developed IC's, and reproduced sound is not impaired in any way as nothing is added artificially. The amplifier section has a music power output IHF of 50 watts at 4 ohms, or 44 watts at 8 ohms with a continuous power of 18-18 watts at 4 ohms or 15-15 watts at 8 ohms. Total harmonic distortion is less than 0.8 percent at rated output, and intermodulation distortion less than 1 percent at rated output. Power bandwidth IHF is 25 to 40,000 Hz; frequency response 20 to 50,000 Hz plus or minus 1 db; channel separation is better than 50 db at 1,000 Hz; hum and noise 80 db; damping factor 70 with 8 ohm load. Block diagram of the QS100 is shown in Fig. 3-11 and a pictorial view is displayed in Fig. 3-12.

The Sansui model QS500, 4-channel rear amplifier-synthesizer-decoder combination is designed to let you step up your present 2-channel system to a true 4-channel overnight and with maximum economy. The powerful 120-watt amplifier offers high performance and the unique 4-channel synthesizer decoder converts 2-channel tapes, discs, and FM stereo broadcasts into the richer 4-channel format. The result closely approximates the vibrant sound fields as they happened at the live performances. A block diagram showing connections to your 2-channel system are shown in Fig. 3-13. Specifications for the power amplifier section are music power IHF 120 watts at 4 ohms, 90 watts at 8 ohms; continuous power 40-40 watts at 4 ohms, 33-33 watts at 8 ohms; total harmonic distortion less

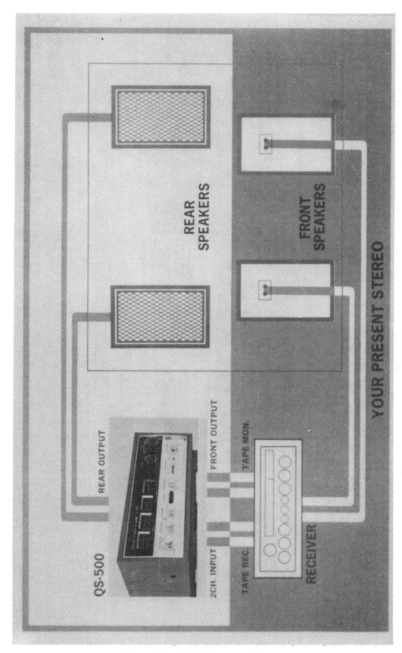

Fig. 3-13. QS-500 connections.

Fig. 3-14. Sansui model QS-500 4-channel converter-amplifier. (Courtesy Sansui Electronics Corp.)

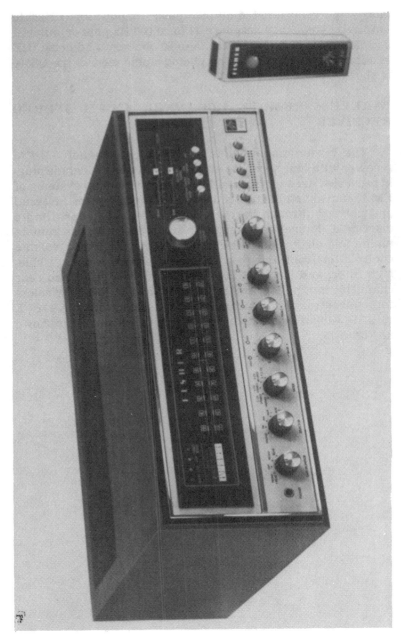

Fig. 3-15. Fisher model 801, 250-watt 4-channel AM-FM stereo receiver-wireless remote. (Courtesy Fisher Radio.)

than 0.5 percent at rated output; power bandwidth IHF 20 to 40,000 Hz; frequency response 20 to 50,000 Hz plus or minus 1 db; channel separation better than 60 db; hum and noise IHF 80 db; damping factor 24. The photographic view of the QS500 is shown in Fig. 3-14.

WIRELESS REMOTE 4-CHANNEL AM-FM STEREO RECEIVER

The Fisher model 801 is a 250 watt 4-channel AM-FM receiver with four separate amplifiers. Features include plug-in multiplex decoder adaptable to future 4-channel systems of FM broadcasting; extracts 4-2-4 recorded program material and plays it through rear channels to closely approximate discrete 4-channel reproduction; automatic wireless remote tuning and electronic touch tuning; separate sliding volume controls for front and rear channels; separate bass, treble, high filter, and loudness contour controls for front and rear channels; muting switch; tape monitor. The WT-50 wireless remote control unit is included. Dimensions are 17" X 5¼" X 16½" D, weight 35 lbs. Pictorial view of the model 801 is shown in Fig. 3-15.

Chapter 4
Stereo Record Players

Recent advances in record players have resulted in numerous improvements in tone arm design, tracking accuracy and the Synchro-Lab motor. The latter represents a true merger of induction motor power plus synchronous motor smoothness to ensure unwavering turntable speed at all times. Since variations in turntable speed do affect the pitch of the recording along with the elapsed time required for playback, it becomes fairly obvious that true quality reproduction meets a major hurdle here. Low speed changes result in wow, while higher speed variations produce flutter. Either wow or flutter are distortion, and this is precisely what we want to avoid in our system.

Although the manual turntable was definitely more popular in the past as a result of its exceptional reliability, the recent major improvements in automatic design have elevated that type of turntable to equal footing. Consecutive play of several discs without attention provides a large plus for the automatic, but fewer parts guarantee greater reliability for the manual. Cartridges have also advanced by giant steps in the last few years.

Requirements for Reliable Performance

The better changers offer several, easy adjustments to enhance the overall performance of the unit, for example, a sliding weight to pinpoint stylus force of the more sensitive phono cartridges. Other adjustments may incorporate magnetic anti-skating controls with separate settings provided for elliptical and radial stylii; manual or automatic options at thirty three and one-third or 45 rpm; multiple play spindles for up to six records; single-play spindle that rotates to minimize record wear; variable pitch speed control for tuning records to synchronize with tape recorder or musical instrument; built-in illuminated prismatic stroboscope for continuous monitoring of speed settings; 15 degree vertical tracking adjustment for manual and automatic play; viscous-damped cueing and pause controls; built-in dust and lint remover to maintain clean record grooves **ahead of stylus**; safety stop to avoid tone arm descent onto plate; and two-point, gentle record support platform.

Many of the more expensive record players provide a built-in stroboscope to make it easy to frequently check turntable speed, and there can be little doubt that this extra expense is well worth the slight additional charge. Although this speed can be checked very simply by viewing an ordinary strobe disc with a fluorescent light, the inconvenience involved would often discourage such checks. The speed should be constant and correct at all times within about 0.2 percent (that is a mere two-tenths of one percent of the proper speed), in order to avoid distortion. The synchronous motor is locked to the line frequency of the AC power, which makes it reasonably easy to keep a constant speed, if other factors are correct. Abnormal friction or drag caused by dust or dirt in record grooves, excessive stylus pressure, or spindle friction may cause variations in speed and such conditions should be corrected. This may be overcome to a degree by additional torque provided in synchronous motors having continuous or multiple poles as well as the hysteresis or Papst type motor used in professional studio record-playing equipment.

OBSERVE TRACKING, ANTI-SKATING, AND OTHER ADJUSTMENTS

Proper tracking must always be maintained with a minimum of stylus force conducive to full recovery of the recorded information. The accurate adjustment arrangement should cover a full range of cartridge and stylus masses. A long tone arm reduces tracking error and makes smooth operation easier to attain. An automatic locking of the tone arm in the rest position may represent a worthwhile investment by eliminating the risk of damage to your valuable recordings, and the stylus as well. The anti-skating system should provide for a separate adjustment for the elliptical as well as the conical stylus as their skating patterns would differ. You may notice that some automatic turntables offer an anti-skating system that moves the tone arm outward as it descends, which could be quite objectionable under certain conditions. The variable pitch control provides a small range of speed adjustment for matching the pitch of a record to another musical source. In looking over the features offered by different manufacturers of turntables in this chapter, you will note the amazing advances in design that have taken place, and although space limitations permit coverage of only a few models, your demands will be better oriented and appreciated when a selection is made as a result of the material covered herein.

ECONOMY UNITS

In a primary source of program material, any economy measures would be improper. This does not mean that anything less than broadcast studio equipment is a waste of money, but rather that a good turntable is good investment in the future of your hi-fi system and one that you will someday be glad you made. Actually, the cost of a good phono unit may run about twice that of an inexpensive variety, and the more you improve the rest of your system, the more obvious the inexpensive turntable will become. The first sacrifice is usually in the quality of the cartridge. It is the cartridge that is going to interpret everything on your records and supply that information to your amplifiers. In reality, only part of the music will ever get through because the inexpensive cartridge just does not have the ability to detect certain portions. Suppose we insist on a higher-priced cartridge for our economy turntable. Even if it fits the tone arm, if it does not have stylus force adjustments, anti-skating and proper tracking, the quality cartridge would be operating far below top efficiency and subject to damage through lack of the usual protective measures. The heavy-duty constant speed motor used on low-priced turntables could not maintain the accurate speeds needed for the high quality reproduction that would be expected from the higher-priced cartridge.

COMPACT SYSTEMS

The space-saving convenience offered by the three-piece stereo has proven popular where area is limited. By com-

Fig. 4-1. KLH model 34 three-piece music system. (Courtesy KLH Research & Development.)

bining the automatic turntable and receiver in a single unit, a small space may accommodate both elements. The two speakers included with such systems may be placed at opposite points in the room for proper stereo coverage. A typical example of this combination is shown in the KLH model Thirty-Four which combines the simplicity of installation, convenience, and size of a three-piece music system with the versatility and high performance level of individual, more expensive and complex high fidelity components. In addition, all the components have been matched to insure optimum performance from each. The control center consists of a special Garrard automatic turntable (made exclusively for KLH), an all-silicon solid-state receiver (FM-FM stereo tuner); tone, volume, balance and switching circuitry; preamplifier; power amplifier and output stages, and two KLH model 32 best-selling speakers. The turntable may be used manually for single play, and automatically with a stack of 7", 10", or 12" records at any of 4 speeds. A low mass tone arm is used with Pickering V-15-AT-2 cartridge matched to the wideband preamplifier. The FM tuner has a usable sensitivity of 2 microvolts, and the amplifiers, a continuous RMS power into 8 ohms of 10-10 watts. See Fig. 4-1.

The BSR model RTS-20A AM-FM phono stereo music system includes the AM-FM-MPX receiver system with 20 watts music power, solid-state design, headphone jack, acoustically matched speakers (pair), walnut grained cabinets and all cables; the model 2000-X Minichanger total turntable with ceramic stereo cartridge, molded ebony base, tinted dust cover, and 45 rpm adaptor. It is offered at a special package price. The RTS-20A is pictured in Fig. 4-2.

The BSR model RTS-29 AM-FM 8-track phono stereo music system includes the AM-FM-MPX receiver system with 20 watts music power, solid-state, 8-track playback, headphone jack, acoustically matched pair of speakers, walnut grained cabinets, with all cables; BSR McDonald 6500-X full size total turntable with deluxe stereo ceramic cartridge, diamond stylus, cue and pause control, anti-skate control, full size turntable, automatic tone arm lock, base, dust cover and 45 rpm adaptor. It is offered at a special package price. RTS-29 is illustrated in Fig. 4-3.

The BSR model RTS-30 AM-FM magnetic phono stereo system includes the BSR McDonald R-30 AM-FM-MPX

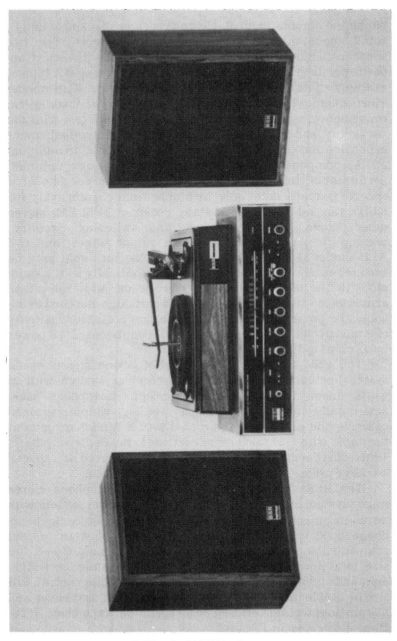

Fig. 4-2. BSR model RTS-20A AM-FM phono stereo music system. (Courtesy BSR (USA) Ltd.)

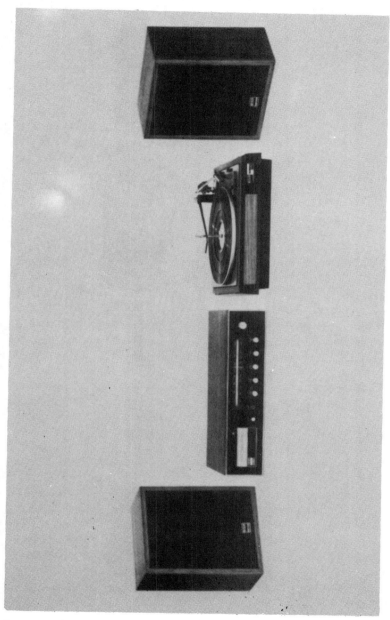

Fig. 4-3. BSR model RTS-29 AM-FM-8-track-phono stereo music system. (Courtesy BSR (USA) Ltd.)

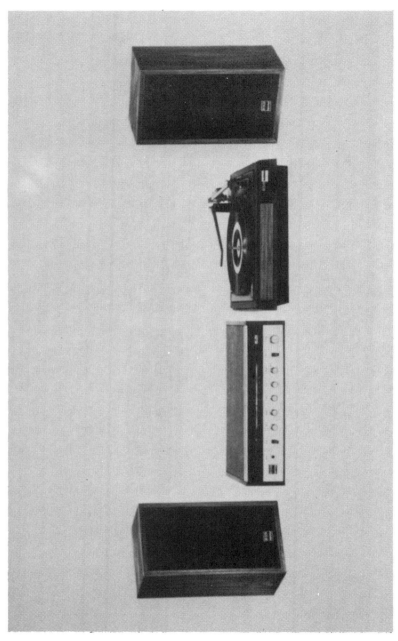

Fig. 4-4. BSR model RTS-30 AM-FM phono stereo music system. (Courtesy BSR (USA) Ltd.)

receiver with 30 watts music power, solid-state, magnetic cartridge input, dual range tuning meter, headphone jack, walnut cabinet; BSR McDonald 310-X total turntable with low mass counter-weighted arm, Shure M-75 magnetic cartridge, cue and pause control, anti-skate control, base, and dust cover; BSR McDonald SS-1 speaker system with acoustically matched pair of speakers in walnut grained enclosures with 12 speaker cables. It is offered as a special package. The RTS-30 is completely illustrated in Fig. 4-4.

Cartridge Trackability

Importance of tracking can hardly be over-estimated. The stylus tip must follow the record groove up to and beyond the cutting limits of modern recording. Not just a certain frequency but the entire audible range must be covered with minimum tracking force.

The Shure V-15, Type II cartridge allows excellent trackability at extremely light force. This cartridge is capable of tracking most records at ¾ gram force, although current advances in recording have produced a growing number of recordings requiring 1 gram of force to recover the required range of recorded information. It should be noted that ¾ gram tracking necessitates a cartridge capable of tracking at that value as well as a high quality manual arm or high quality automatic arm turntable capable of such high values. The practical effect of improved bass trackability may be noted, although you may have been required to increase tracking forces for heavily modulated bass drums, tuba, or even piano passages. But now you may play these passages without increasing your tracking forces and still not witness any bass flutter, or other types of distortion. This allows reduction of 1¼ tracking to 1 gram, or 1 gram to ¾ gram for records with high velocity bass material. Now you may improve your present V-15 Type II with the V-15E Improved stylus, the comparison of which is shown in Fig. 4-5. An enlarged view of the Shure V-15 Type II cartridge is in Fig. 4-6.

The trackability as a meaningful specification is shown clearly by the chart and, unlike the over-simplified and frequently misunderstood design specifications, the trackability is a measure of your overall performance. The chart emphasizes the frequency across the bottom and the

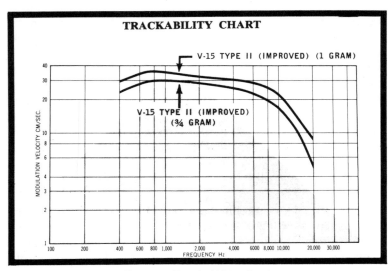

Fig. 4-5. Trackability chart.

modulation velocities on the side, while the gray area represents the maximum theoretical limits for cutting recorded velocities. In actual practice, many records are produced which exceed these limits.

The smoother the curve of an individual cartridge, the greater its distance above the gray area, and the better the trackability. The superiority of the new Shure V-15 Type II Improved from a trackability standpoint is clearly shown by the top solid-black line. As previously emphasized the true high fidelity sound may only be recreated at the source of the sound, and as the camera is no better than its lens, the phono system is no better than its cartridge. This so-called precise, miniaturized, electronic wave generator must carry the full burden of translating the miles of stereo recorded grooves into usable electrical impulses and should do this without adding or subtracting from what is on the recording. The high trackability cartridge consistently and effectively tracks all grooves in current recordings at light, record-saving pressures regardless of the difficult assignment imposed by some instruments. Preservation of fidelity and reduction of distortion from all your records whether old or new is assured.

The Empire 999VE-X long playing stereo cartridge offers a tracking force of ¼ to 1¼ grams and usable output to 40,000 Hz. The hand polished diamond stylus (.2 x .7 mil) ensures extremely low mass. This cartridge is recommended for high performance turntables and changers only and offers similar specifications to the more expensive Empire 1000ZE-X stereo measurement standard. The cartridge is fully shielded, has four poles, four coils, and three magnets (more than any other brand) to produce better balance and better hum rejection. There are no foreign noises, magnetic balance is perfect, and there is a signal-to-noise ratio of 80 db. It features a moving magnetic element and stylus lever system .001 inch thick. The entire cartridge weighs only 7 grams, which is ideal for modern compliance requirements. A simplified sketch of the cartridge is shown in Fig. 4-7. These top of the line cartridges feature high performance parameters designed for 4-channel capability with even greater frequency response and compliance than ever before and track at forces so low they barely touch your records. An enlarged view of the Empire 999VE-X cartridge is shown in Fig. 4-8.

Fig. 4-6. Shure V-15 type II cartridge. (Courtesy Shure Brothers Inc.)

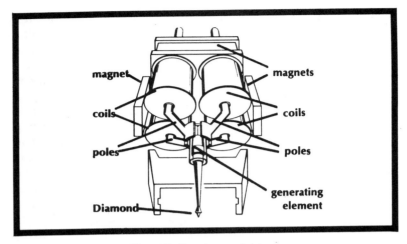

Fig. 4-7. Empire cartridge.

TURNTABLE FEATURES

BSR McDonald model 810 is an automatic turntable which uses a sequential cam system to replace the conventional cam gear and swinging plate used in other automatic turntables. This revolutionary engineering

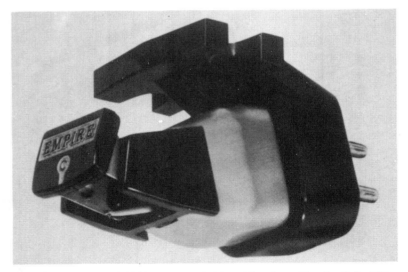

Fig. 4-8. Empire 999VE-X stereo cartridge. (Courtesy Empire Scientific Corp.)

breakthrough results in a smooth, quiet operation and overall reliability. Eight independent, pre-programmed cams eliminate the need for the light stampings and noisy moving parts required in every other automatic mechanism. The longer the tone arm the less the tracking error, and the 8.562 inch pivot-to-stylus length reduces tracking error to less than 0.5 degree per inch. Resiliently mounted, gliding one-piece counter weight provides precise zero balance adjustment over full range of cartridge and stylus masses. Precision micrometer wheel permits continuous infinite stylus pressure setting between 0 and 6.0 grams. The low-mass aluminum tone arm assuring extremely low resonance is counterbalanced in both horizontal and vertical planes. A new, quiet synchronous motor provides unwavering, constant speed that is entirely independent of input voltage or record load. Tone arm automatically locks in the rest position when unit is off to eliminate the risk of accidental damage to stylus or record. It automatically unlocks in any operational function. In many other automatic turntables, the anti-skating system tends to move the tone arm outward while it descends, but the BSR designed cue clutch prevents this. After pause, the 810 tone arm returns to the very same groove every time and the cueing control is operative in both automatic and manual. The same gentle viscous controlled descent provided in the cueing operation also functions during automatic and semi-automatic play. An adjustable, dynamic anti-skate control system provides settings for substantially different requirements of elliptical or conical stylii. The continuously corrected degree of compensation is applied regardless of where the stylus is on the record. Concentric gimbal arm mount is gyroscopically pivoted on four pre-loaded ball-bearing races to assure virtually no friction in either the horizontal or vertical planes, to give a tracking capability of ¼ gram. The stylus overhang adjustment is an important added feature of the slide-in cartridge head, and provision is made on the cartridge slide for plus or minus ⅛ inch range of stylus overhang which can be accurately set by means of the removable locating gauge. Once the stylus overhang is set, the locating gauge can be replaced by a soft stylus whisking brush also provided. Full 12 inch dynamically balanced turntable platter provides maximum record support, and the cast, non-ferrous seven pound platter is machined and precision-balanced for op-

Fig. 4-9. BSR model 810 automatic turntable.

timum performance. Platter mat is deep-ribbed rubber for maximum record protection. The platter-mounted strobe disc enables precise adjustment of turntable speed when pitch control is utilized (both thirty-three and one-third and 45 rpm.) Manual sub-spindle rotates with platter to eliminate center hole wear and interchanges with umbrella-type spindle for automatic play. Variable pitch control provides 6 percent range of speed adjustment, to match pitch of a record to a live instrument or other playback device. The featherweight pushbuttons ensure unexcelled operational flexibility with settings for manual play, semi-automatic play, infinite repeat, or fully automatic play. Jar-free function selection is assured by the pushbutton feature, even under extremely light stylus pressures. It is currently priced at $149.50. See Fig. 4-9.

Thorens model TD-150 Mark II, 2-speed integrated transcription turntable has many of the advanced engineering features and refinements of more sophisticated models. Priced at $140, including tone arm and base, it is a concession only to economics, not to Thorens traditional quality and precision. Other features include precision cueing synchronizer, conveniently positioned at the front, and isolated from the tone arm. This feature suspends and lowers the arm smoothly and gently into the groove, thus extending life of stylus and records. Anti-skate control maintains stylus position precisely on both walls of the stereo groove for distortion-free reproduction. Adjustable low mass plug-in shell permits tracking angle adjustment to ideal 15 degrees. Low friction accommodates high compliance cartridges. Tone arm balance and stylus tracking force regulates with half gram adjustments down to 0.5 gram, with tracking error below 0.2 percent cm of radius. Glide speed adjustment to change speed smoothly without jolting stylus. The precision balanced turntable weighs 7 lbs with 12 inch non-magnetic platter and uses flywheel action to absorb speed variations and eliminate wow and flutter. The unified suspension system prevents rumble effects of the chassis, motor, acoustic feedback and outside vibrations. The double 16-pole synchronous motor provides constant, smooth in-phase speed, regularity better than plus or minus 0.1 percent with drive belt system to further reduce rumble. Speeds are thirty three and one-third and 45 rpm. Dimensions are 15⅝" by 12⅞" by 5"

Thorens model TD-125 series, 3-speed transcription turntable exemplifies the composite of newest engineering concepts, originality in design, attention to detail, and quality components. These no-compromise refinements offer solid-state circuitry replacing conventional mechanical methods for speed control. By reducing the number of moving parts as well as total mass, the rumble producing factor is eliminated. The motor rotor velocity, another rumble originator, is precisely governed by a Wien Bridge transistorized oscillator with frequency varied exactly by a gold-plated switch to change the speed of the motor. Adjustments may be observed instantly on the built-in strobe-scope. Motor speed has been reduced to an incomparable 250 rpm at thirty-three and one-third rpm turntable speed, and in combination with solid-state circuitry and Wien Bridge oscillator, rumble is reduced to an infinitesimal and inaudible —48db (—68 db weighted) or less rumble than any other turntable. A belt-driven 16-pole synchronous motor offers power to spare and the belt-drive system acts as a filter between motor pulley and large diameter flywheel of the turntable. Powered by a push-pull amplifier capable of 20 watts, the motor uses only 5 watts allowing for considerable power to spare. Trouble-free, this miniature dynamo assures constant and smooth in-phase precise speeds at sixteen and two-thirds, thirty three and one third, and 45 rpm. Tonearm and drive system are isolated for shock-free operation with the tonearm and turntable platter mounted on an independent chassis. This unit is shock mounted to another chassis which houses the drive system and controls. Thus the tonearm is protected against shocks when the controls are operated. Acoustic feedback is minimized through the absence of vibrations which are further decreased by the precision polished, self-lubricating bearings of the turntable platter. As the tonearm is off-the-chassis, tonearm and cartridge changeover is simplified and alignment always correct. A full 12 inch (7 lbs.) antimagnetic die cast platter is dynamically balanced for low wow and flutter. All controls are conveniently accessible on front panel. Features include an on-off slide switch, 3-speed selector, fine speed control, illuminated strobe window, and fingertip switch for cueing control.

Garrard model Zero 100, 2-speed automatic turntable represents a truly revolutionary advance in record-playing equipment with an amazing Zero tracking error over entire record surface. The stylus is perpendicularly tangent to the groove throughout the record as a result of the breakthrough in tonearm design combined with other advances to make possible flawless record reproduction. The cartridge housing is pivoted directly above the stylus tip. A Synchro-Lab motor combines induction motor power with the smoothness of a synchronous motor to ensure unwavering turntable speed. A calibrated sliding weight is provided for setting stylus force, tracking the most sensitive cartridges with precision. The magnetic anti-skating control has a separate setting for elliptical and radial stylii. Provides automatic or manual play at thirty three and one-third or 45 rpm with the multiple-play spindle handling up to six records while the single-play spindle rotates with the record to prevent wear. Features plus or minus 3 percent variable pitch speed control for tuning records to synchronize with tape recorder or musical instruments. Built-in, illuminated prismatic stroboscope for continuous speed monitoring. Has 15-inch vertical tracking adjustment for both manual and automatic play with viscous-damped cueing and pause control. Dust and lint remover automatically cleans record groove ahead of stylus. Safety stop prevents tonearm descent onto plate, and the full-size 12-

inch platter has a safety platform for gentle 2-point record support. The unit is priced at $189.50. The Zero 100 is pictured in Figs. 4-10 and 4-11.

United Audio Products model Dual 1219, professional automatic turntable with 3-speeds includes every deluxe feature, with tonearm tracking at less than the optimum recommended force of the best cartridge available. The twin-ring gimbal suspension lets the tonearm pivot like a gyroscope for total freedom and perfect balance in tracking as shown in Fig. 4-12. The special control labeled B in the sketch lets the stylus track at perfect angle in single play and at center of stack in multiple play. Tracking force is applied at pivot (C), maintaining perfect dynamic balance of tonearm. Separate anti-skating calibrations at D for elliptical and conical stylii are provided, as each skates differently. The tonearm counterbalance at E is elastically damped and has vernier adjustment with 0.01 gm. click stops for precision balancing. Pitch control allows tuning records over a 6 percent range and cue control system is damped to prevent bounce. All requirements have

Fig. 4-10. Garrard Zero 100 automatic turntable on WB2 base. (Courtesy British Industries Co.)

Fig. 4-11. Garrard Zero 100 automatic turntable. (Courtesy British Industries Co.)

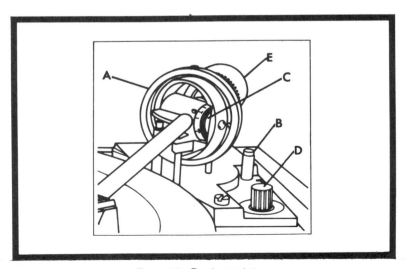

Fig. 4-12. Dual precision.

been met in the past and currently are being exceeded by comfortable margins, and with the four-channel record a reality for all to enjoy, the demands on the tonearm and turntable performance will be much more exacting so the precision of the 1219 ceases to be a luxury, but a necessity. It is priced at $175 less cartridge. The model 1219 automatic turntable combines unrestricted flexibility of operation with complete professional quality, and is actually this company's finest turntable. It offers thrity three and one-third, 45, and 78 rpm plus automatic or manual operation. Handles up to 6 records, with rotating spindle on single play to avoid record wear. The mode selector drops the tone arm for 15 degrees stylus tracking, zero degree vertical tracking error and true friction-free arm movement as a result of the four-point gyroscopic gimbal suspension. Separate anti-skating for conical or elliptical stylii, adjustable stylus overhang and 12-inch turntable includes a dynamically-balanced 7-pound platter. The synchronous, continuous pole motor combines

Fig. 4-13. Dual 1219 automatic turntable with WB-19X base & DC-9X dust cover. (Courtesy United Audio Products, Inc.)

Fig. 4-14. Dual 1218 automatic turntable. (Courtesy United Audio Products, Inc.)

high torque with absolute speed accuracy regardless of line voltage changes. The cue control system is damped to prevent bounce, and counter-balance is elastically damped with precision balancing through 0.01 gram click stops. A pitch control permits record tuning to other external music sources through 6 percent range. Pictorial view is shown in Fig. 4-13.

United Audio Product's Dual 1218 automatic professional turntable with its precision features and fine performance has become the most popular this company has ever made. Most of the features of the more expensive 1219 are included, with the twin-ring gyroscopic gimbal to center and balance the tone arm within both axes of movement. Features include flawless tracking as low as ½ gram; high torque synchronous motor; cueing damped up and down; anti-skating scales separately calibrated for conical and elliptical stylii; compact chassis 11" x 13", price only $155. Pictorial view of the 1218 is shown in Fig. 4-14.

The AR turntable, pictured in Fig. 4-15 is guaranteed to meet NAB specifications for broadcast equipment on wow, flutter, rumble, and speed accuracy. This means 0.1 percent maximum wow and flutter combined when measured as specified. The speed accuracy required is plus or minus 0.3 percent, or 21 lines per minute drift on a standard 216 line stroboscope card. All AR turntables are tested to meet this standard, and remain within limits using an extra 5 grams stylus force which would be equivalent to the drag of a dust bug. Line voltage variations have no effect on the speed of the turntable with its synchronous drive motor.

Empire Troubador model 598 turntable is designed exclusively for the new low tracking force cartridges that will not wear out your records. Features include the Empire 990 playback arm with a friction measurement of 1 milligram; arm fully stereo balanced; sealed instrument ball-bearing races; stylus force dialed with calibrated clock mainspring (dials 0 - 4 grams plus or minus 0.1 gram); exclusive Dyna Lift which automatically lifts arm off record as music ends; micrometer anti-skating adjustment for conical or elliptical stylii to eliminate uneven record or stylus wear; arm counterweight zero balance adjustment 2-8 grams; in-line stylus-to-pivot axis (no warp, wow or cosine-error distortion); 5-wire circuit eliminates ground loops, plug in installation; hysteresis synchronous motor (self-cooling, high torque with inside-out rotor) reaches full speed in less than one third of a revolution while locking in on AC line frequency—zero error; built-in strobe disc and pitch control; flexible belt drive precision ground to plus or minus .0001 tolerance; 12-inch

Fig. 4-15. AR turntable. (Courtesy Acoustic Research, Inc.)

Fig. 4-16. Empire Troubador model 598 turntable system. (Courtesy Empire Scientific Corp.)

turntable platter with 3-inch thick balanced drive flywheel; microhoned oilite bearings and lapped chrome steel shafts machined as individually matched pairs; rumble 90 db, wow and flutter .01 percent; customized record mat holds records by outer rim, grooves never touch mat; pneumatic suspension, tracks as low as 0.1 gram; dead center cueing control; 3 speeds (thirty three and one third, 45, 78 rpm); pushbutton power control; built-in 45 rpm spindle; overall dimensions with base and dust cover, 17½" W by 15⅛" D by 8" H; Swiss ground gold finish, priced at $199.95 less base and cover. See Fig. 4-16.

The Dual model CS16 integrated module is a new type turntable as illustrated and features a dynamically-balanced low-mass tone arm that tracks flawlessly as low as 0.75 gram. Uses constant-speed high-torque motor with pitch control to compensate for off-pitch records and silicone-damped feathertouch cue control. Other extras include elevator-action changer spindle and 12" dynamically balanced platter. Automatic or manual operation at thirty three and one-third, 45, and 78 rpm with multiple play spindle holding up to 6 records. Complete with base, Shure cartridge, and dust cover, it is priced at $119.50. The CS16 is illustrated in Fig. 4-17.

The Elac-Miracord model 625 incorporates many traditional Elac-Miracord features including light touch pushbutton operation; precise cueing; effective anti-skate; balanced four-pole asynchronous motor, and balanced tone arm. It is priced under $100. The model 625 is illustrated in Fig. 4-18.

The Lenco model L85 professional turntable features precisely controlled speed and pitch; viscous damped spring mounting; and electronic shut-off and arm lift. All disadvantages of the mechanical shut-off which causes distortion during the last third of a record due to the mechanical loading of the tone arm, are eliminated by the unique electronic system. The speed can be varied for pitch control by plus or minus 3 percent, while return to precise designated speed is easily accomplished through the built-in illuminated stroboscope. The chassis is furnished with 4 viscous damped

Fig. 4-17. Dual model CS16 integrated module. (Courtesy United Audio Products, Inc.)

Fig. 4-18. Elac-Miracord model 625 changer. (Courtesy Benjamin Electronic Sound Corp.)

suspension springs which permit the unit to be leveled perfectly. Adjustable damping pillars are slotted so that each of the suspension springs can be adjusted exactly by the simple use of a coin. Specifications include speeds of thirty three and one third and 45 rpm with electronic fine speed adjustment plus or minus 3 percent; wow & flutter, weighted plus or minus 0.08 percent; rumble, weighted —63 db; rumble, unweighted —45 db; motor, 16-pole synchronous with belt drive; stylus force adjustment, calibrated 0-5 grams; dimensions: chassis 16¾" x 12⅞", with base 18⅛" x 14⅜" x 3"; power requirements, 12 watts, 110V 60 Hz. The L85 turntable is pictured in Fig. 4-19.

The Elac-Miracord model 770H automatic turntable shares all the important exclusive features of the 50H model such as hysteresis synchronous motor, external stylus overhang adjustment with built-in gauge, massive dynamically balanced turntable and cueing in both manual and automatic modes. Added features include TRU-TRACK, an adjustable head that can be set so that the cartridge assumes the precise 15 degree vertical angle for any number of records when used automatically, or for a single record when used manually. Variable speed control has digital stroboscopic speed indicator. Digital readouts are always visible on the rim of the turntable. Another plus is the built-in ionic elapsed time stylus wear indicator which keeps tabs, by the hour, of precisely how long your stylus has been in use. It even reminds you to check your stylus. The 770H is priced at $225 less cartridge and base. Refer to Fig. 4-20 for pictorial view.

The Elac-Miracord model 50H Mark II automatic turntable includes all the features in the original 50H plus new flexibility and precision in speed control. When requiring pitch control for musical or other purposes, the speed of the 50H II can be varied over a 6 percent range for a semi-tone of pitch

Fig. 4-19. Lenco model L85 professional turntable. (Courtesy Lenco Div., Benjamin Electronic Sound Corp.)

Fig. 4-20. Elac-Miracord model 770H automatic turntable on base WB-700. (Courtesy Benjamin Electronic Sound Corp.)

adjustment and can be restored to precisely accurate speed in an instant with the built-in, illuminated stroboscopic speed indicator. Without this indicator, a pitch control would have you listening more often at the wrong speed than the right one. Whenever you want to interrupt a record momentarily, one smooth touch on the cueing lever floats the arm up gently from the groove; you can float it down again to the same groove at another fingertip command. Or you can position the arm, in raised position, anywhere on the record. Miracord cueing disengages the arm totally during play, so it cannot cause drag. The hysteresis synchronous motor is the type relied on by professional sound engineers to maintain locked-in speed accuracy even in the face of extra record loads or wide fluctuations in power-line voltages. Specifications include speeds of thirty three and one-third, 45, and 78 rpm; rumble —40 db (NAB); wow 0.06 percent; flutter 0.02 percent; tracking error less than 0.4 degree per inch; stylus force adjustment calibrated 0-6 ½ gm. accurate to within 0.1 gm; lateral arm bearing friction, less than 0.05 gm. Chassis size is 14⅛" W, 12⅛" D. Power requirements are 20 watts at 117V AC, 60 Hz. See Fig. 4-21 for illustrated view of the 50H Mark II.

Fig. 4-21. Elac-Miracord model 50H Mark II automatic turntable. (Courtesy Benjamin Electronic Sound Corp.)

NEW RECORD PLAYING SYSTEM DEVELOPMENTS

The photoelectronic cartridge offers a new concept in phono techniques by reducing the work load of the stylus. Since it is only required to modulate a beam of light, its task has been lowered to minor proportions while interpreting the musical information on the record grooves. The signal energy comes from the external power supply, rather than the movement of the stylus in a magnetic field. Thus the cartridge is so free to follow the recorded detail that frequency response is exceptionally flat all the way from 20 through 40,000 Hz.

Toshiba model SR-50, light-sound record playing system utilizes this new development in phonograph performance and design. Since its signal is not generated by the work the delicate record groove performs upon the stylus as in conventional magnetic cartridges, the elliptical stylus of the SR-50 moves a tiny precision shutter to modulate a weightless light beam. This not only ensures less record wear, but also provides the flat frequency response enjoyed. The extremely low distortion level is 0.65 percent, and the output from the phototransistor detectors in the cartridge head is fed to an all-silicon FET preamplifier-power supply in the base. You can bypass your amplifier's phono preamp which is usually the noisiest part of your system, and a tuner, tape or auxiliary input. This reduces noise and permits use of your present phono input with an auxiliary record changer. The preamp contained in the base also provides electronically-controlled optimum stylus force and anti-skating gauge with direct readout based on cartridge performance rather than gram valuation. The end of the record is sensed by photoelectronic means which raises the tone arm and shuts off the turntable automatically. Anti-skating and viscous-damped cueing are included. A feedback network pinpoints the servomatic DC motor speed at precisely thirty three and one-third or 45 rpm regardless of line frequency or voltage. Double-floating suspension-isolation does away with feedback and floor-shake problems.

Toshiba model SR-40E, integrated circuit record playing system uses the C-300 IC cartridge with a specially designed IC in the cartridge to make it possible to use an ultra-miniature ceramic generating element of high compliance and low mass. The performance surpasses the moving coil types in flat response, reduced distortion and excellent transient characteristics. The microminiaturization of the entire cartridge, including built-in IC amplifier-impedance converter, reduces the size to about that of a matchstick. So tone arm mass is correspondingly reduced along with tracking force. Tubular construction of the tone arm eliminates resonance, and displays a tracking error of only plus or minus 1.5 degrees. The cueing device raises or lowers the arm gently at any point on the record surface or with automatic lift at the end, returns the arm to rest position and shuts off the turntable. The turntable platter is an aluminum die-casting and is driven by a vibration isolating belt off the hysteresis synchronous motor with locked-in speed despite power fluctuations. The double floating system has the arm and turntable mounted on a floating subpanel, isolated from both motor and the turntable mounting board, to offer additional protection against vibration. The greater isolation from motor noises thus provided virtually eliminates distortion from acoustic feedback and normal floor shake. The system is enclosed in deluxe, wood grain finish base with conventional plastic hinged dustcover.

Chapter 5

Tape Recorders

Cassette type units with the portable battery-AC capability are useful for business and entertainment purposes. They are inexpensive and the batteries are normally obtainable just about everywhere. While better quality microphones, VU meters, remote and other accessory devices will add to the normally low price of these handy types, such extras will frequently be worthwhile.

A four track stereo cassette recorder including portable battery-AC operation, detachable swing out speakers, twin VU record level meters, separate slide volume and tone controls, instant-on solid-state circuitry may run very close to a hundred dollars. Many of these use the ever popular D cells which offer long life in such use as well as being low in cost. Recording while driving with the attached mike is very useful and since hands, feet, and eyes are free, there is no safety hazard involved even if someone accuses you of talking to yourself.

A more elaborate cassette type stereo recording-playback system may be about $200 with switchable bias, standard-CrO_2, automatic shutoff, mike and input record level controls, sound with sound mixing, main or remote input jacks, input for record changer, tuner, or tape deck; semi-slot tape loading, and matching acoustic suspension speaker systems. It records and plays back 4-track stereo and 2-track monaural. Features include push-button tape controls, pause button for cueing and editing with front panel left and right channel mike inputs, tape recorder output jacks.

A solid-state stereo 8-track cartridge recorder-player tape deck is available at $139.95 with the capability of recording 8-track cartridges through your stereo amplifier, receiver, or live mikes. Record features include stop after each channel 1 through 4 and eject, sound with sound mixing, pushbutton manual eject, pushbutton fast-forward, left and right channel microphone input jacks on front panel, record safety interlock, illuminated track indicator lights, dual recording VU level meter, separate dual mike, and auxiliary volume controls. Playback features include pre-select and play any one channel and stop, automatic eject, pushbutton track selector, and continuous or repeat play. The automatic cartridge ejector protects the tape head and cartridge from damage with power off, and fast-forward allows quick selection of tape sections.

The 2- and 4-channel stereo 8-track record-playback tape deck at about $250 permits recording and playback of 4-channel or 8-track stereo tape cartridges through your 4-channel stereo receiver or amplifier. Two channel, 8-track tape cartridges may also be recorded or played back. Numerous features are included such as, 4 VU record meters, 4-channel (four track) stereo recording and playback heads, 4-track erase, 4-channel demo tape, connecting cables and 4 preamplifiers.

The Fisher TX-420 stereo master control amplifier with built-in 8-track cartridge player and stereo 4-channel converter is shown in Fig. 5-1. The 2+2 decoder extracts hidden ambience information from conventional stereo programs and feeds it into the rear channels of the TX-420, which amplifies the rear channels of a 4-channel system. It recovers encoded 4-2-4 material and the classical-popular switch enhances sound quality. Output is 50 watts. Dimensions are $16\frac{3}{8}$" by $4\frac{3}{4}$" by $11\frac{3}{4}$" deep.

IMPORTANCE OF MECHANISM DRIVE

It is essential to maintain constant motor speed along with good starting power in order to avoid the wow and flutter problems which result in distortion and lack of true sound reproduction. The high torque enables quick action when the function switch is depressed and this instant action must be available in either direction. Any tendency to drag as indicated in some units, shows insufficient power to meet the need and is not acceptable. If several inches are required before attaining full speed, a lack of adequate torque in the motor may be the reason and this means trouble, so be sure to avoid this. When making your selection of a tape machine, be sure that the mechanism drive has the power to do the job with

Fig. 5-1. Fisher model TX-420 stereo master control amplifier. (Courtesy Fisher Radio.)

some margin to spare and also operates at a constant speed regardless of line voltage variations. An insignificant deficiency in the beginning can only get worse. Do not take that gamble here.

CASSETTE, REEL-TO-REEL, OR CARTRIDGE (8-TRACK)

Any preference of one of the three types of tape machines over the others can only be contingent on the personal likes and dislikes of the user. Each type is better than the others in some specific way, and at the same time, not as good in other respects. You may want to settle for two types of machines later on, but the selection of the first one must be made after weighing the question of how it is to be used in your home. If the main purpose will be listening to prerecorded material, and a simple compact unit is preferable, the 8-track cartridge player is the answer to your needs. Plenty of good tapes are available no matter what your musical dictates may be and the simplicity and compact arrangement of these units are hard to beat. In addition, four channel sound may be enjoyed with the cartridge system where the amplifiers and speakers for the mode are included in your system.

Anyone desiring to record for fun with family, friends, speeches, dictation, etc. with a minimum of fuss should look seriously toward the cassette. This type of machine has the answers to compactness, simplicity, and good quality reproduction too for home recording and playback. Good music is also available for listening pleasure if desired, but the portable feature makes the machine perfect for all occasions on vacation, during business trips in the car, motel, or factory for recording notes, speeches, dictation, or any material of interest for later use. The tapes are inexpensive and recording requires no special preparation or skill, and quality is still very good. Battery operation often proves quite useful, and when the AC receptacle is handy, a flick of the switch connects that source of power to the cassette machine. When driving between calls, clip the mike to the sun visor, dash, or coat and prepare for the next stop on cassette tape. No attention is required as the tape stops automatically at the end or when the mike switch is released, in the meantime, both hands are free for driving, and you can record your notes, reports, observations, etc. as you roll along. The fidelity of the reel to reel was better than the cassette at one time, but with the

Fig. 5-2. Panasonic RS-736US.

tremendous advances of the latter, this is no longer true. Special quality tapes are now available for the cassette and its fidelity is second to none.

Those desiring that special professional-like touch have only one choice: the reel-to-reel tape machine. This complicated recording method is made to order for them. Editing, cutting, and all kinds of special effects provide unlimited possibilities for the enjoyment of those interested in such recording sessions. The reel-to-reel, or open reel machine as this type of tape player-recorder is known, offers a choice of speeds by providing 7½ inches per second (ips) for high fidelity recording and/or listening, as much of the commercial tape is at this speed; 3¾ ips for voice or background music;

and even 1⅞ ips which is satisfactory for speech only. The higher the speed of the tape the better the quality or fidelity of the material will be, but the shorter the time to run. Most of the machines require considerable space, and especially with the average accessories offered. Prices start at close to $150 and continue on up to $750 or more for top professional quality. Some of the better open reel machines will be reviewed later in this chapter, such as the Panasonic RS-736US Tape Deck in Fig. 5-2.

ACCESSIBILITY AND CONTROL FEATURES

Check for easy access to the heads for routine cleaning and degaussing. Covers should be secured by slides or snap-fits and should pull off with light pressure to expose the heads. If it is necessary to take things apart to get to these vital components including pressure rollers and guideposts, you may have a tendency to skip the simple cleaning chore which could lead to more serious problems later. Look over the controls provided; are they adequate to cover all conditions that may arise as to functions, modes, levels, etc. Some models exclude those needed less frequently in order to allow a price saving, but if you ever need that specific control or jack or connection for accessories that is not available on your model, the small saving could prove quite expensive in the long run. Sometimes it pays to make provisions for extras that may be added to your system in the future.

LEVEL METERS

The importance of the level meter becomes apparent while struggling to maintain the proper recording level or amplitude without one. How can overmodulation and the distortion that results be avoided otherwise? Getting the exact amplitude for satisfactory recording fidelity may be quite a chore when blindfolded by lack of a meter. Even the inexpensive neon-bulb indicator proved useful by eliminating the risk of providing too much for the recorder to handle, but the advent of solid-state amplifiers ruled out such indicators which required much higher voltages than those allowed in the modern units. The VU meter is the most valuable type and these are normally included on the panel of most tape machines, and actually enable the user to maintain any desired level of modulation according to the recording conditions involved. So the advanced transistorized techniques that have added greatly to the overall reliability of our hi-fi systems, have made it necessary to measure our modulation with a meter as it should be done.

BATTERY-AC OPERATION

Most of the less expensive cassette tape player-recorders do provide battery or AC operation, and this is handy for the average person who wishes to take his recorder along in the car, boat, or any place where AC is not close by. Compact, sturdy, and versatile along with being inexpensive the battery-AC powered unit provides the mixing of business with pleasure by permitting notes, reports, or other business operations to be carried out at the convenience of the owner, and with a minimum of expense. Thus, the low-priced tape cassette machine is ideal for a beginning in the tape field, and if not for business and fun, then certainly for the valuable experience gained which will be most helpful later as the professional-quality tape machine becomes more than a dream. The open reel tape machines are not available in any degree in the battery-AC power source arrangement as their size and power needs are hardly conducive to portability. However, if your requirements dictate such a dual purpose in this type, a few are available but not at the bottom of the price field.

TAPE PROBLEMS AND ANSWERS

The efforts of the cassette manufacturers to provide a higher fidelity machine in addition to the Dolby system of noise reduction have improved the naturalness of the performance through narrower gaps in the head. The recording head is, after all, a miniature horse-shoe magnet with recording activity taking place between the pole pieces. The narrower the gap, the more precise the recording will be and the harder magnetic material recently developed also makes it possible to provide a better head than was available only on the higher priced open reel machines previously. Developments in tape have advanced considerably as only iron oxide was formerly available—tiny particles of rust coated onto a plastic strip. It was found that by processing the iron particles in a particular way, more of them could be attached to a tape. This meant that better frequency response and less background noise was possible, but the recorders required a higher bias current. Chromium dioxide was introduced by Dupont with similar properties, as a substitute for iron oxide. The better cassette decks have bias adjustment switches to make it possible to switch between tapes with the new formulation for high fidelity recording of music and standard formulation cassettes for less critical recording. Cobalt alloys as well as other tape materials are finding increased use. This will continue in the future and the variable bias on the better cassette machines will become even more of a necessity. The acceptance of the cassette has been so great and widespread that improvements in the tapes as well as the machines can be expected to continue to exhibit the concentrated efforts and abilities of the industry in major breakthroughs in the future as in the past.

SIMPLE CLEANING AND DEGAUSSING

After removing the plastic cover, the heads are exposed and should be easy to clean. A Q-tip dipped in ordinary alcohol used for rubbing may be used. Or, if one of the better TV tuner cleaners is available, spray the end of the Q-tip with that and gently rub across the head face, drying with the other end of the tip. You should find rust colored material on the end of your improvised cleaning tool. These oxide particles stick to the heads and accumulate until the tape no longer contacts the head and sound quality suffers. This cleaning job is so simple that you should do it every few weeks or as often as you see a noticeable accumulation on the end of your tip.

Residual magnetism builds up from regular use and causes the noise level to increase until it becomes quite annoying. A degaussing instrument is quite inexpensive and available at most hi-fi stores. Simple to use and very effective, the tool will demagnetize the heads in a fraction of a minute. Plug-in the degausser about five feet or so away from the tape machine, and bring it closer gradually until the tip of the tool is about an inch from the head. Go over each head slowly along with associated metal parts, guides, and posts—not quite touching each with the degausser. Then back away slowly

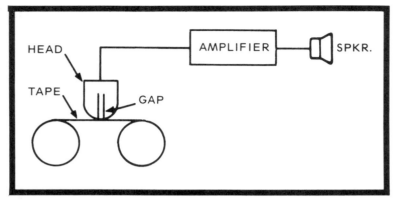

Fig. 5-3. Tape player block diagram.

several feet before turning the tool off or unplugging it. This completes the operation, the metal molecules have been shaken out of their magnetized formation by the strong 60 Hz magnetic field of the degaussing tool.

The head contacting the tape, as previously mentioned, is shown in the simplified block diagram of the playback in Fig. 5-3. The recording process may use the same head and amplifier with a microphone in place of the speaker. Pushbutton switches change the circuit accordingly. Many of the more expensive machines use separate heads for recording and playback as certain advantages are gained as will be noted in the description of these later.

PUSHBUTTON VALUE

Quick, foolproof, and convenient function selection is offered on most cassette machines with the handy pushbutton feature. Fast stops, backups, and replays are thus provided to enable the user to study, review, or alter the recording quickly and accurately. If you want to recheck your report, speech or what have you, simply push the stop button and follow with rewind for instant action. After the rewind function has returned the tape to the desired point, push to stop, then to play, and you get an immediate replay of the part desired. If that section needs changing for any reason, just push stop, rewind, stop, record (interlock) and deliver the change into your microphone, and follow with another stop, rewind, stop, play for an instant check of the change you made. This type of quick action is only possible with pushbutton control, and no other arrangement will allow you to gain the full advantage of this convenient cassette player-recorder. Do not consider any other system of function control for your machine. If you intend to do much recording, insist on the fast-forward (F-F) function too, not included on some models.

DIRECT RECORDING FROM AM-FM RECEIVER

Recording from the receiver is normally a simple procedure, and merely requires cables to be connected between the receiver tape output and tape recorder input or mic (microphone) terminals. Such cables are available at all hi-fi shops, TV appliance stores, etc. and are inexpensive. They must have the proper type of plug on each end to fit your receiver tape output and the recorder input, but this should present no problem as all combinations are offered. Do not have the cables any longer than is necessary as restrictions are involved here. If you are monitoring the recording you are aware of whether everything is working properly, but if you are not, check back on the recorded tape after making your connection initially to make sure that receiver is being recorded. Sometimes the packaged cables are defective and this will either prevent the recording material from getting through to the tape, or distort it to an unacceptable extent. If this happens, disconnect the cables and return them for replacement. If the shop was too far away or even out of town, you may get a friend to check it for you to make sure before returning it for credit. Some tape cassette type recorders include battery-AC portable AM-FM receiver.

The KLH model 21 professional FM radio is a complete high performance monaural receiver which may energize an external speaker, supply a signal directly to a tape recorder for top quality recording of FM broadcasts, or serve as a tuner of superb quality for a separate sound system. The internal speaker may be switched off as desired while listening to an accessory speaker across the room. The solid-state circuitry provides performance equal to that of expensive component receivers, and excellent sensitivity plus selectivity help bring in the most difficult stations free of distortion or noise. No antenna is required in most city or suburban locations, but for fringe area reception, 300-ohm antenna terminals are

Fig. 5-4. KLH professional FM radio model 21. (Courtesy KLH Research & Development Corp.)

provided. The vernier tuning knob and planetary dial allow quick, easy, and consistently accurate tuning, refer to Fig. 5-4 for pictorial view.

Craig model 2402 stereo tape recorder with auto-reverse, records and plays in both directions. Features include sound-on-sound and sound-with-sound recording, detachable speaker enclosures with hi-compliance speakers, three automatic modes (one way, return, repeat); convenient input and output jacks on side panel, tape-break switch and sensing poles, pinch roller released at auto-stop or tape break, three-speed selection by slide-level control. Currently priced at $299.95. Specifications are reel size 7" max.; recording time 12 hours stereo, 24 hours mono; rewind 1200 feet in less than 2 minutes; tape speeds $1\frac{7}{8}$, $3\frac{3}{4}$, $7\frac{1}{2}$ ips; wow and flutter less than 0.15 percent RMS at $7\frac{1}{2}$ ips, less than 0.25 percent RMS at $3\frac{3}{4}$ ips; output power 16 watts total (8-8); S-N ratio better than 40 db; frequency response 30 Hz to 18,000 Hz at $7\frac{1}{2}$ ips; separation better than 40 db; cross-talk 60 db; 14 transistors and 6 diodes; record system quarter-track, 80 kHz bias; erase system AC 80 kHz; inputs (2) microphone, 50K —70 dbm, (2) auxiliary, 220K —20 dbm; outputs (2) speaker, 8 ohms, (2) line, 1K, 0.5 volt, (1) stereo headphone, 8 ohms; speaker size (2) 4-inch hi-compliance; microphone (2) dynamic, omnidirectional 10K;

power 117V AC, 60 Hz, 90 watts; dimensions 19½"W, 15"H, 12½" D; accessories included, one 7 inch take-up reel, two microphones, four patch cords, sensing tape, two detachable speakers.

Kenwood model KW-4066 stereo tape deck (3 head) is designed for the connoisseur of fine music and fine stereo equipment. Dependability is incorporated in a new, exclusive tape transport system featuring an auto-stop mechanism built into the head housing. The pre-amplifier uses low-noise silicon transistors for a high S-N ratio and low distortion. Slide-type level controls, as used in modern broadcast studio equipment, make it easy for you to produce top quality recordings. Tape monitor switch provides a fast, easy method of checking sound during recording process, mode switch produces natural monaural sound from two-track tapes, and easy-to-read large illuminated dual VU meters are included. Specifications are 3 heads (playback, recording, erase); speeds 7½ and 3¾ ips for hi-fi recording and 1⅞ for speech; frequency response 25 Hz to 12,000 Hz at 3¾ ips, 25 Hz to 6,000 Hz at 1⅞ ips; S-N ratio better than 50 db; wow and flutter less than 0.15 percent at 7½ ips, less than 0.25 percent at 3¾ ips, less than 0.35 percent at 1⅞ ips; inputs, mike-line; outputs, line-headphone jack; dimensions 16" W, 7⅛" H, 12¾" D. Accessories include extra 7 inch reel, 2 connecting cords, 2 reel caps, 2 reel adjusting discs.

Kenwood model KW-5066 stereo tape deck (4 head) provides every professional recording capability including an advanced design recording bias adjustment system. A built-in test signal oscillator generates a reference signal to allow proper control recording bias for optimum fidelity. In addition to the usual heads for recording, playback, and erase, a full-track erase head is provided. Sound-on-sound and echo controls facilitate a wide range of creative recording. A noise filter eliminates undesirable hiss on pre-recorded tapes. Separate input and output level controls permit easy monitoring, and front panel dubbing connector is provided. Specifications are frequency response 25 to 20,000 Hz at 7½ ips, 25 to 12,000 Hz at 3¾ ips; S-N ratio better than 50 db at 7½ ips; wow and flutter less than 0.15 percent at 7½ ips; inputs; mike-line; outputs: line-headphone jack; dimensions 16" W, 7" H, 15½" D. Accessories supplied are extra 7 inch reel, 2 connecting cords, 2 reel caps, 2 reel adjusting discs.

Fig. 5-5. Revox model A77 Mark III stereo tape deck. (Courtesy Revox Corp.)

Kenwood model KX-7010 stereo cassette deck provides economy and convenience along with frequency response expected only from conventional reel-to-reel equipment. Quality and dependability are assured when professional results are desired in cassette recording. Other features include micro-gap heads to provide superior sound reproduction, low-noise silicon transistors in preamplifier, easy-to-use pushbutton controls for all tape modes, pop-up tape housing for easy cassette loading and unloading, pause control for instant stop or start, high filter for elimination of hiss noise on recorded tapes, professional slide-type level control, dual illuminated VU meters, 3-digit automatic counter with reset button, front panel headphone jack for monitoring. Specifications are 4-track stereo-mono recording and playback; S-N ratio better than 45 db; wow-flutter less than 0.2 percent. Size is 10½" W, 4" H, 9" D. Weight is 7 lbs.

Revox model A77 Mark III stereo tape deck has all the time-tested features of earlier models plus those changes which would meaningfully improve performance and reliability. The new oscillator circuit is designed for greater efficiency and lower distortion. The self-adjusting braking system has been modified and strengthened, and a new hardening process devised to reduce capstan wear. Improvements in tape handling and spooling have been incorporated plus numerous significant minor changes. The new A77 Mark III is pictured in Fig. 5-5.

The Kenwood KX-700 model stereo cassette deck with the Dolby noise reduction system combines reel-to-reel quality with cassette convenience. Its patented Dolby (Type B) system virtually eliminates high frequency background noise without affecting high frequency signals. The super-ferrite record-play head, with precision micro-gap, utilizes bias frequencies to optimum advantage for quality reproduction. A three-way tape selector permits a choice of correct bias for regular, low-noise or chromium dioxide tapes. Special features, in addition to the aforementioned, include slide lever volume controls, two large VU meters, three digit counter, piano key controls, input-output terminals, low distortion preamp with high S-N ratio, and automatic stop. Specifications are: heads—4-track, 2 channel stereo and monaural Super Ferrite Core—Record-Play and erase; double drive, hysteresis synchronous motor; regular, low-noise, chromium dioxide tapes, C-30, C-60, C-90, C-120; tape speed 1⅞ ips; wow-flutter less than 0.13 percent; frequency response 25-15 kHz(chromium dioxide tape),25-13 kHz (regular tape); S-N with Dolby 55 db regular tape, 58 db chromium dioxide tape; without Dolby figures are 45 db and 48 db respectively; recording bias 85 kHz; separation between track or channels, more than 40 db. This model is illustrated in Fig. 5-6.

Harman-Kardon model CAD4 stereo tape cassette deck has been engineered expressly for the serious recording enthusiast, and includes professional features and performance to compare favorably with most reel-to-reel recorders. A

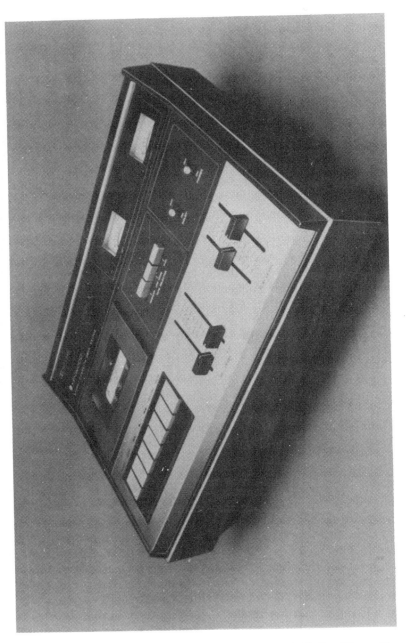

Fig. 5-6. Kenwood KX-700 stereo cassette deck with Dolby (Type B). (Courtesy Kenwood.)

push-pull bias oscillator ensures minimum noise and distortion while the newly designed record-playback head employs four laminations per stack rather than the usual three found in conventional cassette tape decks. This permits reproduction of a much broader frequency range extending from 30 to 12,500 Hz, and the new pole piece reduces contour effect that is responsible for excessive bass boost or lack of clarity in the lows. Two large, illuminated VU meters permit all types of material to be recorded accurately and without distortion. The overmodulation light immediately indicates tape overload at +2VU on either channel. Precise standards in the manufacture of the tape handling mechanism ensure smooth tape transport as capstan drive, heavy duty motor, tape guide, and clutches are carefully calibrated. The electronic speed control holds variations to a minimum and subsequent wow and flutter as well. Other features include automatic digital counter with pushbutton reset, all silicon low-noise solid-state devices, pushbutton function switches, automatic tape cassette ejection switch, record interlock that prevents accidental erasing, instant on, motor on light, unique electronic automatic shutoff, recording indicator light, two separate level controls, mono-stereo microphone switch and two mike inputs, power on-off switch with indicator light, inputs for low and high level program sources, solid steel construction which provides rigidity and mechanical alignment of moving parts. Specifications are: tape speed 1⅞ ips; harmonic distortion less than 1.5 percent; speed variation below 2 percent; wow and flutter 0.25 percent; frequency response 30 to 12,500 Hz; S-N ratio better than 49 db; crosstalk better than 35 db; erasure better than 55 db; high-level sensitivity 200 mv, plus or minus 2 db for Zero VU; high-level input impedance 200K; low-level input sensitivity 0.2 mv, plus or minus 2 db for Zero VU; low-level input impedance 2.5K; output level 0.8 RMS plus or minus 2 db at maximum recording level. Dimensions are 12½"W, 3¼"H, 9"D. Weight is 10 lbs.

The Wollensak model 2516AV assures convenient, efficient, and reliable operation with many additional features. It offers a frequency response of 60 through 12,000 Hz at 1⅞ ips with wow and flutter 0.25 percent RMS. Signal-to-noise ratio exceeds 46 db and power output is 8 watts per channel EIA. Dimensions are 13⅞" by 9¼" by 4½". Weight is 20 lbs. Features, and inputs and outputs are shown in the layout of

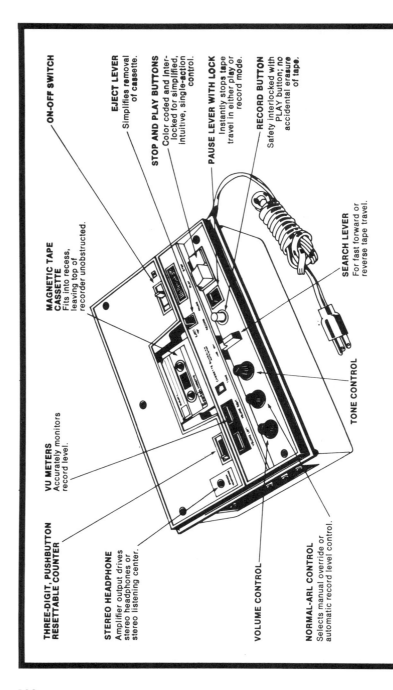

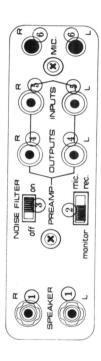

① **SPEAKER OUTPUTS:** Jacks for connecting output of recorder's playback amplifiers to external speakers.

② **MONITOR SWITCH:** Controls output of both channel's SPEAKER jacks, but only during recording.

③ **NOISE FILTER SWITCH:** Activates a tuned filter network to reduce hiss from poor tapes.

④ **PREAMP OUTPUTS:** Jacks for connecting recorder's playback preamplifiers to an auxiliary amplifier.

⑤ **PREAMP INPUTS:** Jacks for recording from hi-level outputs of hi-fi amplifier, tuner, radio/phono, other tape recorder.

⑥ **MIC INPUTS:** Jacks for microphone connection.

Fig. 5-7. Model 2516AV cassette layout features (Wollensak).

Fig. 5-7. This cassette deck also includes tone control and eject lever along with resettable counter.

The TEAC model A-24 stereo cassette deck offers a dual record-level meter which clearly indicates the signal level during recordings as well as during playback. Specifications are tape speed 1⅞ ips; wow and flutter 0.2 percent; S-N ratio 45 db; frequency response 40 to 12,000 Hz; dimensions 4¼" H, 13⅝" W, 9⅝" D; weight 11 lbs. Power requirements are 115V AC, 20 watts. It is priced at $179.50.

KLH model 41 stereo tape deck features open reel system with Dolby and utilizes 3 heads. Wide-range, noise-free recordings are assured through the Dolby noise reduction system and a signal-to-noise ratio at 3¾ ips 66 db, at 7½ ips 68 db. This unit features four track stereo record-playback; two record VU level meters; 4-digit tape counter; single tape function control for rewind, stop, play, pause, and fast forward; separate erase, record, and playback heads; dimensions 5⅜" H, 14¼" W, 11⅜" D. It is priced at $199.95.

Harman-Kardon model CAD5, professional tape cassette recorder with built-in Dolby noise reduction processor provides an incredible boost in bandwidth and dynamic range. No other concept in recent years has been so well received by audio experts as this famous noise reduction process, and the limits of signal-to-noise range have been broadened by as much as 15 db. The CAD5 offers features, specifications, and performance at least equal to the most advanced and versatile professional reel-to-reel recorders. Inherent problems within the cassette system, such as narrow, thin tape and the small separation between tracks, have formerly made low noise figures impossible. Improvements in the tapes with new uniform oxide and high saturation types have shown considerable advances in the S-N ratio, but the noise problem although lessened, still was there. However, the 10 db improvement in the S-N ratio resulting from the addition of the Dolby B system enables the user to wipe out tape hiss and other objectionable though unidentifiable background to present a clean, natural sound. The system boosts low-level, high frequency signals that were down in or near the noise level before recording, but does not change the high-level signal. During playback, these same boosted signals are attenuated or reduced to their precise original level and at the same time noise is reduced to a level below audibility. This is completed without any perceptible effect on the signal. Other features of the CAD5 include the newly designed deep-gap head with four laminations per stack to provide extended frequency response with bass to 30 Hz showing unprecedented linearity; accurate digital counter with pushbutton reset; modular construction ensures complete uniformity and reliability; balanced capstan drive with advanced guide mechanism design for precise handling of tape; pushbutton switches for operating functions; record interlock eliminates accidental erasure; Dolby in-out switch; front panel indicator light for motor; two illuminated VU meters; sliding record level controls; inputs for low and high level sources; pushbutton pause switch; heavy duty motor for fast tape transport and low wow or flutter; and overmodulation indicator light to prevent recording distortion. Specifications are tape speed 1⅞ ips; total harmonic distortion (THD) less than 1.5 percent wow and flutter 0.15 RMS at 1⅞ ips; frequency response plus or minus 2 db below 30 to beyond 12,500 Hz; signal-to-noise better than 55 db (weighted below zero VU); bias oscillator 105 kHz; crosstalk 40 db; record-playback output 1 volt; erasure 55 db; high level input sensitivity 200 mv plus or minus 2 db for zero VU; high level input impedance 200K. Dimensions are 3¼" H, 12½" W, 9" D; Weight is 10 pounds. See Fig. 5-8, showing pictorial view of model CAD5. The controls and layout detail include the counter with reset above cassette compartment, pushbutton function switches below, dual VU meters to the right, off-on power switch extreme right with the Dolby processor in-out beside. The sliding record level controls are just left of the Dolby switch and continuing left in order Dolby indicator, overmodulation indicator, motor light, record light, and stereo-mono switch for microphones.

The Sansui model SC-700 Dolby system stereo cassette tape deck features the answer to the claimed inability of the cassette tape deck to reproduce true hi-fi sound. Hardly in the inexpensive category, the SC-700 does measure up to the other components in the quality stereo system, and for a compact means of enjoying true high fidelity sound, this component is well worth the investment. A precision instrument able to make or reproduce 4-track stereo or two track monaural recordings, this ultra convenient model incorporates every major advance made to date in the design and manufacture of first class cassette tape decks. Features include 3-digit tape

Fig. 5-8. Harman-Kardon CAD5 tape cassette deck with Dolby system. (Courtesy Harman-Kardon.)

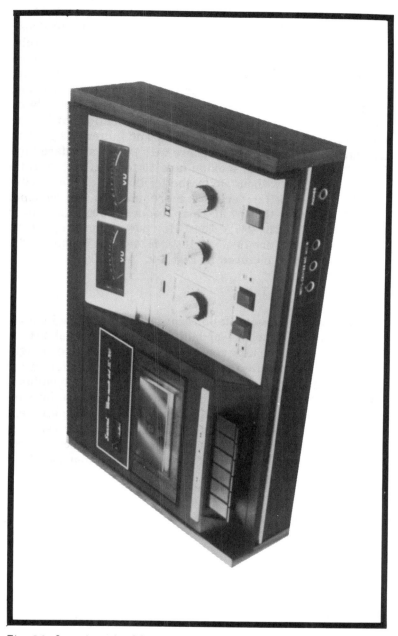
Fig. 5-9. Sansui model SC-700 stereo cassette deck with Dolby system. (Courtesy Sansui Electronics Corp.)

counter, pause mechanism, Dolby anti-hiss switch, tape selector switch, DC servo motor, automatic shut-off-release mechanism, large VU meters, dual recording level controls, record-play preamp, high recording bias frequency, contourless record-play head, optional output circuit voltage, easy dubbing arrangement, dual output level controls, headphone jack, 3-mike mixing, computer-type gold-plated terminals, optional extras available. Specifications are: tape speed accuracy, within 1 percent; wow and flutter 0.12 percent weighted RMS; frequency response with Dolby, standard tape 40 to 13 kHz, chromium dioxide tape 50 to 16 kHz; S-N ratio better than 50 db without Dolby, better than 56 db with Dolby (58 db above 4 kHz); erasure more than 60 db; channel separation, better than 47 db; bias frequency 100 kHz; power voltage 100, 117, 220, 240 V 50-60 Hz; power consumption 23 watts. Dimensions are 15¼" W, 10⅛" H, 4⅛" D. Weight is 12.6 lbs. The SC-700 is pictured in Fig. 5-9.

Use of Chromium Tape

Some of the cassette tape recorders are equipped with a switch to permit the use of both standard or regular tape (ferro-oxide) and the tape coated with chrome-dioxide, a high-performance magnetic material. The latter type offers a far higher capability to retain magnetism than the conventional tape, and has been used for high-density recordings in computerized equipment. Since it is several times more efficient than standard tape in recording high frequencies, it requires that a tape selector switch be used **without fail** whenever used. This will ensure recordings of a higher signal-to-noise (S-N) ratio and far less distortion.

Chapter 6
Connectors & Cables

The antenna requirements for the AM-FM receiver are usually adequately filled by the built-in or included arrangement, and it seldom is necessary to install more sophisticated types. However, more effective antennas will be needed in some areas for sufficient FM pickup, especially if normal reception is desired from certain stations in your nearby area. The usual method of using the house wiring to carry the FM signals to your tuner is not always efficient enough, as the network of wires may not be compatible to your individual needs. Other metallic materials in their proximity or even between them and the FM station involved can reduce their signal pickup to such a low level that it is useless. Many TV antennas include special FM elements that provide an ideal antenna for your FM as well as your TV. If you have a signal splitter at your TV set, as many do, an FM connection may already be available there, otherwise you may pick one up at your local TV-stereo store along with enough wire to reach the FM antenna terminals on your receiver. In some cases it may be advantageous to have a small FM type antenna in the attic, or for apartments, outside the window installations should be considered where permissible. Make sure that the antenna is pointing in the direction offering the best signal from your favorite station, as you will notice that orientation makes quite a difference. Where twin leads are used at the antenna terminals, some improvement may be noted on weaker stations if the leads are reversed, and if only a single wire is connected to a terminal screw, try it on the second terminal for a possible increase in signal strength. If noise or other types of interference become annoying, a bandpass filter for FM should be connected between the antenna lead-in and the set.

CONNECTING ACCESSORIES

Connecting accessories to your stereo system should not prove difficult as cable connectors and adapters are readily available to cover just about any condition normally imaginable. There will be some specific requirements, according to the purpose of the cable, regarding number of wires, size of wires, and whether shielding is necessary. These needs will be answered for you if you are using a connector made for the indicated purpose. However, if it becomes necessary to improvise, certain procedures will be mandatory. We will go into some such cases. Just as certain connectors were necessary on the ends of your cables in order to fit the point where they were connected, the impedance (opposition to alternating voltage) of the cables must be considered when making certain connections. These impedances must match if we are to get the maximum transfer of energy from one cable or wire to the other. In the speaker system, the matching formula becomes quite apparent as we take a look at Fig. 6-1. Our amplifier output terminals call for 8 ohms, so the speaker or speakers must add up to exactly 8 ohms to ensure an impedance match. This is the only way to get maximum energy from our amplifier into our speaker, and if only one speaker is used, an 8 ohm one will serve nicely. Suppose four 8 ohm speakers are connected to the amplifier and they must add up to 8 ohms to match our 8 ohm amplifier terminals. Connecting as shown in Fig. 6-1 is quite simple and matches perfectly. Let us examine alternatives. Connecting the four speakers across the amplifier terminals is called parallel or shunt and would add to 2 ohms for the speakers across the 8 ohm amplifier output. This would deplete the output and cause a tremendous loss of energy in the transfer. Connecting the speakers in a string (series), with the two ends of the string connected to the output terminals, would provide 32 ohms which is a mismatch and also results in poor power transfer. The series-parallel arrangement of the speakers shown in Fig. 6-1 is the only correct answer.

How do you know whether a connection is parallel or series? A parallel connection always provides additional paths. One speaker connected to the amplifier offers a single 8-ohm path of impedance or opposition to the flow of power from the amplifier. When a second 8-ohm speaker is connected in

138

parallel (across) with the first, two paths are available for current flow, so the opposition is halved or divided by two. (4 ohms equals 8 divided by 2.) Adding a third speaker in parallel, we have three paths and the opposition is further reduced to 8 divided by 3 or 2.67 ohms.

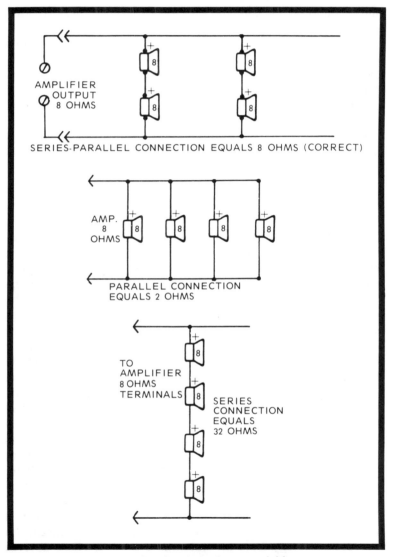

Fig. 6-1. Speaker impedance matching.

What about series connections? A series connection extends the path of opposition as power from the amplifier must pass through one speaker to reach the other. Since the path is extended, the opposition of the first speaker must be added to that of the second speaker, so two 8-ohm speakers would offer 16 ohms. By the same token, when three 8-ohm speakers are connected in series, we have a single path three speakers long and merely add to get the total opposition of 24 ohms.

Series-parallel represents a combination of the two methods of connection, and the fundamental rules still apply. Refer to the series-parallel connection of Fig. 6-1 where two strings of two 8-ohm series connected speakers are connected in parallel. The series connections are each 8 + 8 or 16 ohms and two 16-ohm strings in parallel provide two 16-ohm paths, or 16 divided by 2 equals 8 ohms. This simple method could get complex if unequal values were used but this would not be practical in speaker circuits.

IMPORTANCE OF CONNECTING CABLES

The variety of connecting cables available and the importance of selecting the proper types for the job can hardly be over-emphasized. Most types are readily picked up at your neighborhood electronic or TV shop. The connection to the antenna, where one is required, for your FM tuner, should be made with the correct type in order to avoid unnecessary loss of signal. When additional FM stations are desired, more antenna facilities will be needed, and certain steps must be followed to get the most signal from the outside FM antenna to the FM tuner input terminals. Avoid using lamp cord, zip cord, etc. for lead wire. Stick to conventional lead-in wire as suggested for your tuner. In most cases 300-ohm twin-lead will do nicely and any of the popular brands will be fine. A good quality here is well worth the difference, it lasts longer and works better (less loss) too. Some localities have severe electrical interference problems, and a shielded type of lead-in will reduce or even eliminate these nuisances. The use of coaxial cable (75 ohms) will work very well here and if the 72-75 ohm input terminal is not provided on your FM receiver, a matching transformer should be used to get maximum transfer of energy from the antenna to the set. These devices are inexpensive and are carried in your local hi-fi store. Even though performance may be satisfactory without such extras, there will be conditions under which this will not be true, so do not economize here! The recommended coaxial cable is RG-59U, and when this type of shielded lead wire is needed, it should be requested. The connecting cables from the tuner to an external amplifier should be a shielded type, but unlike the coaxial cable used for antenna lead wire, the matching problem no longer exists in the cable. We are not interested in the ohmic value here, only in the number of conductors and the quality of the shielding. There will be cases where hum in the system can be traced to these amplifier cables, resulting from loose connectors or shields connected at both ends producing a ground loop. These will be discussed in detail later.

GETTING MAXIMUM OUTPUT TO SPEAKER SYSTEMS

The main purpose of any power amplifier is to deliver full (natural) power to the loudspeaker system which represents the load. We have taken care of the sound reproduction quality all through our hi-fi system and by high gain amplifiers have magnified that tiny electrical wave from the source until it is big enough to drive the largest speaker system our pocketbook could provide. The power amplifier has made this big sound available. Now it must be transferred to the speaker system without loss or distortion. For greatest power transfer from a power source to a loudspeaker (load), the impedance of the former must be equal to the latter. Impedance, simply stated, is resistance or opposition to the fluctuating voltage representing our music to this point. When the output impedance of the amplifier is equal to the input impedance of our speaker system, they are said to be matched, and we will get maximum power into our speaker system. The matching problem has been taken care of by the manufacturer of your amplifier, up to the output terminals that are to be connected to the speaker system. Often, allowances are made at this point to provide for optional systems having different loudspeaker impedances as 4 ohm, 8 ohm, and possibly 16 ohm. By connecting your speakers to the appropriate terminals, a match is assured. Inside our amplifier, we have an arrangement like that shown in the sketch of Fig. 6-2. The output transistor is matched to the primary winding of the transformer and that power is inductively coupled into the other winding (secondary) of that transformer, much the

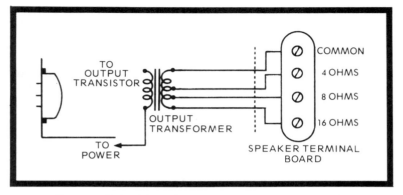

Fig. 6-2. Output transformer connections to terminals.

same as your home receives its power from the high-voltage lines topping the telephone pole through the pole transformer. The secondary winding matches the speaker and where more than one choice is available, the winding has been tapped to accommodate each value. These should be used to provide the exact match necessary for the specific speaker system being connected, and when adding additional speakers the value overall will change. Examples of multiple speakers connected to add up to the required value for the amplifier output terminals are shown in Fig. 6-1. Speakers connected in series, plus (+) terminal of one speaker to negative (—) or common terminal of the next etc. are known as series connected and their impedance in ohms add. Three 4-ohm speakers connected in series would offer a total impedance of 12 ohms. Connecting these same speakers in parallel, (+) of one to (+) of the next and on to the third speaker (+) terminal, and then connecting all three (—) or common terminals together would give us 4 divided by 3 ohms total impedance for the parallel connection. Now we could take 9 speakers and connect them in 3 groups of 3 using parallel connections for each group and each would have a four thirds ohm impedance total, but connecting the 3 groups of three in series, the total of the 9 speakers is 4 ohms again as this is a series-parallel connection. Thus it becomes obvious that we can easily arrive at most any necessary total impedance for our speaker system by proper connections, and this enables us to match the output impedance of the amplifier. There are several other ways of matching and these will be covered more in the chapter on speakers.

Fig. 6-3. AM-FM receiver (rear).

BASIC TERMINAL LAYOUTS

The basic terminal arrangement of the AM-FM stereo receiver showing speaker terminals for left and right channels along with tape input and tape output terminals are illustrated in the photo of Fig. 6-3. Simple 300-ohm antenna terminals are also included. Another type of antenna terminal arrangement is shown in Fig. 6-4 which provides for either type of antenna connection for the FM, regular 300-ohm twin-lead or the 75-ohm coaxial cable (RG-59U) often used in noisy areas. Provisions are made for an additional AM antenna to be used along with the usual internal one included inside most receiver cabinets. Output terminals for connection to your speaker system may be the simple type shown in Fig. 6-3 mentioned above or may either include two or three ohmic values as in Fig. 6-2. Plug-in arrangements are provided in most of the later units, and these would appear like Fig. 6-5 showing the basic 2-channel stereo and the 4-channel (left-right-front-rear). Basic connections between the ordinary stereo output to the 4-channel decoder and amplifier for rear speakers is

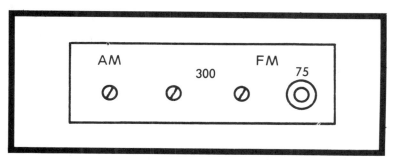

Fig. 6-4. Antenna terminals for either type FM lead-in.

143

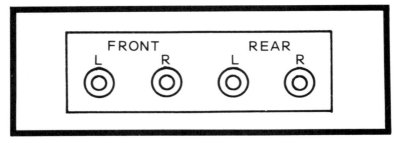

Fig. 6-5. Plug-in 2-channel or 4-channel.

pictured in Fig. 6-6. This shows how easy it is to connect this inexpensive unit to your system and evaluate the new surround sound right in your own home. Then if you like what you hear at that point, there is plenty of room to add to your system from there. Many roads for improvement will be available. Some decoder units include the rear channel amplifier while others either suggest adding it externally or even bypassing it. The 4-channel sound is discussed thoroughly in Chapter 3.

INSTALLING EXTRA SPEAKERS

Speakers may be added to the system very easily by sticking to the rules for matching them to the output of your amplifiers. We must come up with the exact total ohmic value

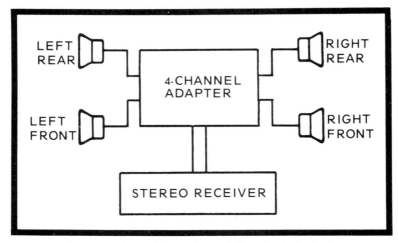

Fig. 6-6. Connecting 4-channel decoder (adapter).

as previously discussed in this chapter. If the speakers are to be added in other rooms, or outside recreation areas such as patio, porch, swimming pool or yard, you may use a switching arrangement (built-in on some amplifiers) or a permanent connection can be selected. The switching unit may be purchased at most hi-fi stores, and is recommended in cases where several areas are to be covered. Any matching needed will be taken care of in the unit itself in most cases, but if in doubt, ask the sales person. The permanent connection may be made directly to the amplifier through a series resistor to match the new load to the amplifier. Ask for a non-inductive resistor of the value needed for your particular installation. A typical connection of this type is shown in Fig. 6-7, but your system could require a different value if the speakers involved are other than those shown. Do not forget to observe phasing of the speakers, (+) to (+) for parallel connections and (+) to (—) or C for series connections. In many cases, only the (+) terminal of the speaker is marked as there are only two terminals. Ordinary zip cord (2 conductor) wire is satisfactory for the average home speaker being added unless it is necessary to come near neon lighting or any other high-voltage wires. Shielded 2-conductor wire should be used to avoid picking up interference noise in such cases. Where exposed to the weather in outside installations, the speaker wires should be the outdoor type to avoid frequent replacement.

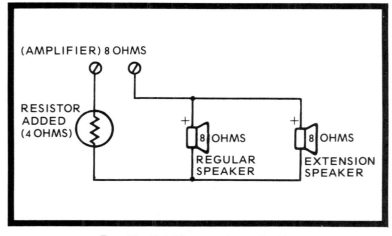

Fig. 6-7. Matching speaker extension.

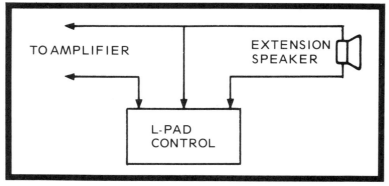

Fig. 6-8. Typical L-pad volume control.

EXTERNAL LEVEL CONTROLS

The commonly used control for external level adjustment on your extra or remote speakers is the L-pad which maintains a constant load on the source regardless of the volume level on the speaker being controlled. These are available in several types with various connections; some have screw terminals and others must be soldered. Another type has two jacks for plugging in extension speakers with convenient screw type terminals for connection to amplifiers. This makes an ideal control for a second set of speakers, and should be 8 ohms when used with 8-ohm speakers. Most L-pad controls have an 8-ohm value, but may be found in 4- or 16-ohm values and these are required with certain speaker systems of the same value. The remote speaker fader control provides volume adjustment and may be used to increase the level on one speaker while reducing it on another. A speaker selector switch for solid-state stereo systems is a useful addition for easy selection of one of three pairs of speakers and uses non-shorting contacts capable of handling up to 60 watts of audio. The T-pad may also be used for adjusting speaker volume and is a constant impedance control usually having a value of 8 ohms. Several conventional arrangements for connecting these controls may be used; one is shown in Fig. 6-8.

PLUGGING IN STEREO HEADPHONES

The extra hours of listening pleasure that may readily be offered through a set of stereo headphones makes this a welcome addition to any good hi-fi system. Although the overall quality can hardly equal that of a good speaker system, the enjoyment is sufficient to ensure the continued interest of most. This form of private listening with its complete freedom from disturbance to any other activities in the vicinity, enables the enthusiast to indulge in many periods of relaxed pleasure that would otherwise either be denied or inconvenient. Most stereo receivers have a headphone connection (jack) available on the front panel and others may offer a similar connection for headphones in the rear of the receiver-amplifier. As the headphones come with a stereo phone plug attached, all that is required here is to insert the plug into your headphone jack and adjust the volume level. The price range of stereo headphones is from slightly under $20 to over $100, and the best method of selection is to try them for comfort, weight, and performance. The high-frequency response is important and should be carefully checked to satisfy individual taste; any early cutoff on the high-end will undoubtedly become annoying in the long run. Accessories are often used with stereo headphones to afford convenience of remote operation with coiled extension cords and control boxes. These may provide connections for one or two sets of headphones with volume controls for each channel plus speaker on-off control switch.

PRIVATE LISTENING FEATURES

Several makes and models of stereophones are currently offered and quite a wide price range is available from which to select those of your choice. Features and some specifications are listed on a few of the types to assist in a selection according to individual requirements. The KLH model 80 stereophones are shown in use by the photo of Fig. 6-9.

Koss model PRO-4AA stereophones offer smooth, fatigue-free listening about two octaves beyond conventional dynamic types. The wide headband is adjustable and cushioned for maximum comfort with patented, liquid-filled, removable cushions. It is designed for use with all stereo hi-fi amplifiers from one to 500 watts. The high quality drivers are mounted on acoustically designed enclosures to ensure smooth frequency response from 10 to 20,000 Hz. A coiled 10 ft. cord and plug are included in the $60.00 price.

Fig. 6-9. 9 KLH model 80 stereophones. (Courtesy KLH Research & Development Corp.)

Koss model ESP-6 electrostatic stereophones offer an entirely new concept in headphones with self-energized diaphragms which reproduce three octaves of sound beyond the limits of voice coil and cone type driven headphones. Frequency response is 30 to 19,000 Hz plus or minus 2 db and usable response 10 to 20,000 Hz. Features include input impedance of 4 to 16 ohms; cushioned foam rubber headband for maximum comfort plus liquid-filled ear cushions to minimize annoying external noise, pilot lamp indicators on each ear piece show sound level, 10 ft. coiled cord, T-3 connector box with speaker on-off switch, and carry case. They are priced at $95.00.

The stereophones of today provide exceptional fidelity for private listening and as a result are enjoying a popularity exceeding that of grandpop's era of crystal sets and one-tubers. The range of selection is extensive, both pricewise and as to the degree of sophistication, with some ranging over $100. The two major types, each represented above, are the dynamic, which operate through diaphragm activity in a magnetic field, and the electrostatic, by diaphragm movement in an electrostatic field. The latter type have exceptionally sensitive reproduction capabilities extending through all ten octaves, and are used in most professional sound applications where precise monitoring is mandatory. The four-channel sound trend has caused headphones to be introduced to adequately permit the audiophile to experience this full-dimensional sound privately, such as through the new Koss K 2+2.

Koss model K 2+2 quadrafones feature two-channel stereo or four-channel stereo with a flick of the switch. This new stereophone permits full enjoyment from either two or four channel equipment and the four channel mode energizes four separate driver elements in the K 2+2 to surround you with the exciting quadraphonic sound. Additional features include volume controls at the base of each ear cup for adjusting the front drivers. This feature makes it possible for you to move your position from front row center to the middle of the concert hall without getting out of your easy chair. They are currently available at $85.00 in your favorite hi-fi shop.

Superex model QT-4 discrete four-channel stereo headphones are designed for use with 4-channel stereo receivers or amplifiers. The two acoustically matched transducers in each earpiece are electrically independent of each other. Frequency response 20 to 18,000 Hz and impedance of 4 to 16 ohms with fully adjustable padded headband. The urethane vinyl covered foam ear cushion is replaceable. A 15 ft. coiled cord with 2 stereo plugs for front and rear is included at the $50.00 price figure.

Pioneer model QT-4 discrete four-channel stereo headphones are designed for use with 4-channel stereo receivers or amplifiers. The two acoustically matched transducers in each earpiece are electrically independent of each other. Frequency response 20 to 18,000 Hz and impedance of 4 to 16 ohms with fully adjustable padded headband. The urethane vinyl covered foam ear cushion is replaceable. A 15 ft. coiled cord with 2 stereo plugs for front and rear is included at the $50.00 price figure.

Pioneer model SE-50 stereophones feature two transducers in each earpiece with individual tone and volume controls. Reasonably flat response is provided over the entire range. They are sold complete with 16 ft. coiled cord and priced at $49.95.

Stanton model 770 electrostatic stereophones feature very wide range and smooth response. Separate energizer is included which connects to speaker amplifier terminals, and 10 ft. coiled cord. Very light weight and comfortable, they are priced at $150.

Lafayette model F-990 stereophones feature large 3½" transducers for wide frequency response with good coverage from 20 through 20,000 Hz., foam-filled earphone cushions and adjustable, cushioned vinyl headband. Complete with stereo phone plug and 6½ ft. cord, the price is $29.95.

Sennheiser model HD-414 stereophones feature lightweight, compact, open air design. Featuring high impedance (2000 ohms), good high end response and balance, with 10 ft. cable and stereo phone plug, they are priced at $33.95.

Panasonic model RF-60 stereophones with built-in FM-FM stereo radio operate from 3 AA batteries and sensitive antennas. Tone control switch is provided for instant selection of preferred bass or treble. Cost complete is $75.00.

Telex model Studio 1 stereophones are designed for professional use and offer a range of response from 20 to 22,000 Hz. Each phone is equipped with slide type tone and volume controls. Maximum input power is one watt with impedance of 3 to 16 ohms. Supplied with 25 ft. of coiled cord. Price is $99.95.

Beyer model DT-480 stereophones feature extreme comfort and frequency response of 30 through 18,000 Hz with rapid drop off below 100 Hz, but excellent overall sound quality. Impedance 5, 25, or 200 ohms and maximum input power 200 milliwatts. Supplied with 7 ft. cord. Price is $75.

Sharp model 770 stereophones feature liquid-filled removable ear cushions which provide maximum isolation. Frequency response is 20 to 20,000 Hz and maximum input one watt. Impedance 4 to 16 ohms. Supplied with 14 ft. cord. Price is $100.

PLUGS AND JACKS

The matter of plugs on the ends of connecting cables, as well as the matching jacks into which they must be inserted may appear to present a headache to the average system builder. It is indeed fortunate for most of us that the packaged connectors came along to offer an acceptable answer for all but a few odd varieties. Even though defects do exist in some of this material, getting a connecting cable with plug A on one end and plug B on the other is certainly appreciated. Standardization in electronics is something we have always wanted but unfortunately never achieved. If you have the information as to the make and model, or at least the make, of the units to be connected, most sales personnel are able to help you select the cables with the proper plugs on the ends to fill your specific needs. There are also numerous adapters for changing from one type of plug or jack to another, for example you have a phono plug to connect to a standard phone jack. The adapter makes it simple, as it has a phono jack input with a standard phone plug output which takes care of the situation very nicely. Adapters to fit any desired combination are offered, even one to couple phono plugs together by accepting a phono plug at each end. If you cannot decide what you may need in the way of connecting cables, the universal cable kit containing an assortment is a good buy at less than $3.00. DIN adapter cables are available for European recorders to match the standard RCA type phono jacks and plugs, so do not give up. No matter how sticky your situation appears, there is a solution. The plug and jack are often referred to as male and female respectively in audio circles, in fact the latter terms will be used in all phases of electronics. In-line connectors and speaker connections for Norelco, Wollensak, Bell & Howell, Telefunken and many other European units using 2-pin, 3-pin, and 5-pin DIN types are available (Weltron).

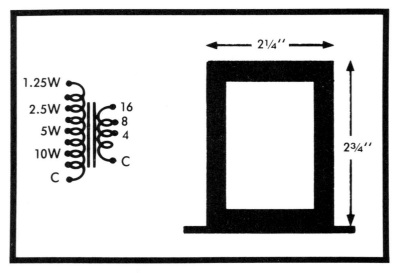

Fig. 6-10. Constant voltage transformer.

CONSTANT VOLTAGE LINES

When several extension speakers are needed the constant voltage line is often the best solution to a rather difficult situation from a standpoint of power requirements. For example if six 8-ohm speakers were needed on a single extension line, connecting the six in parallel would drop the total impedance to 1.33 ohms which would be much too low for any system to handle without overheating and probable damage to the amplifier. The mismatch would be so great that power transfer efficiency would drop to less than 10 percent. The constant voltage line offers the answer and with a few small CV transformers, it is quite simple. Most power amplifiers have a constant voltage output (70.7 volts or 25 volts) and a line from these terminals may be run to the most distant speaker. The transformer in each case should be located at the speaker with the transformer secondary connected to the voice coil from common to one side of voice coil and from the other side of voice coil to the proper tap on the CV transformer as shown in Fig. 6-10. The primary of the CV transformer is connected across the 70 V or 25 V line from the amplifier. It is as simple as that. Each speaker is connected according to its impedance, and the primary of each transformer is connected at the tap for the amount of power needed at that location. This

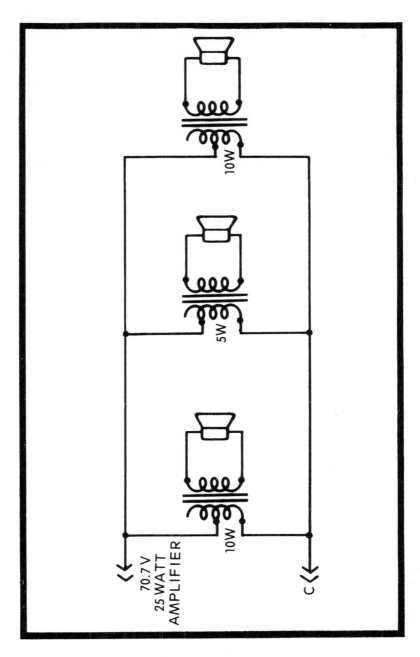

Fig. 6-11. CV line-transformer connections.

is shown in the simple sketch of Fig. 6-11. Note that the power level on any speaker in the line may be varied according to the taps on its transformer. This system works somewhat like the AC power line to your home except that voltage and frequency are not constant here as they would be in the power line. Nevertheless, the same output voltage is maintained regardless of the number of speakers connected as long as each has its own CV transformer and it is within its rated capacity. If our amplifier has a power output of 100 watts per channel, we must stay within those limits, in other words 10 ten watt transformers or 5 ten watt and 10 five watt. Our total demand must not exceed the rated capacity of our amplifier.

Hi-Fi Speaker Systems

Chapter 7

Hi-fi speaker systems, although at one time offering little to the quality of the hi-fi chain, have made such giant steps in recent years that we must concede that a share in the overall glory in sound reproduction belongs to them. An inexpensive speaker using the 2-speaker 2-way acoustic suspension system may run as low as $15 in a fiberglass walnut veneer enclosure. The 5" woofer and 3½" tweeter handle 20 watts with an overall response of 60 to 19,000 Hz. The 3-speaker 3-way acoustic suspension system handles 40 watts of program material with changing speaker characteristics according to the type of material reproduced. Retailing at $90, the system affords a frequency response of 20 to 20,000 Hz using a 10" woofer, a 3" cone-type driver, and a 1½" super-tweeter with aluminum cone.

Anyone desiring to create their own speaker installations and system arrangements may select unmounted speakers according to their specific requirements. These quality hi-fi speakers usually are available for less than $10, but may run around $25-30 for greater power handling, say to about 50 watts.

The Sony SS-4200 speaker system offers a unique, full-range 3-way system combining 8" woofer, with 8" mid-range driver and two 3" dome-cone tweeters in a fully sealed, 3-way acoustic system. The oiled walnut enclosure is 23¼" high, 13¾" wide, and 11⅞" deep. The Sony SS-9500 Omni-Radial speaker system provides six full-range speakers equidistantly spaced and radially mounted around the perimeter of the enclosure. This projects the sound 360 degrees with the capacity to handle the most powerful amplifiers. Walnut finish enclosure is 23⅝" high, and 16" in diameter at mid point.

SB-750, Panasonic's 3-way floor type speaker system, uses a sealed air suspension, 5-speaker system. The input impedance is 8 ohms with a peak power handling capacity of 85 watts and a frequency response of 20 to 20,000 Hz. A 12-inch rolled polyurethane edge high compliance woofer, two 6-inch craft-pulp dome radiators for mid-ranges, and two 4-inch duralumin dome radiators for tweeters are used. Front mounted level controls handle mid-range and tweeter adjustment. Features include multi-channel or full-range terminals and switch, detachable base, lattice grill, and finished speaker-mounting board. Teak wood cabinet measures 30" H, 19¼" W, 11¾" D.

Marantz model Imperial 7, 3-way bookshelf system utilizes a 12" woofer, 3½" mid-range, 1¾" tweeter. Special features include multi-position high-frequency control to permit you to fine-tune for individual room acoustics to suit your personal taste. Specifications are: frequency response plus or minus 5 db, 40 Hz to 20,000 Hz; frequency range 30 to 20,000 Hz; power rating 100 watts integrated program material; controls: 3-position high-frequency level selector switch; 3-position mid-range level selector switch; nominal impedance 8 ohms; dimensions: 25½" H, 14¼" W, 11½" D.

Dynaco model A-50, aperiodic speaker system minimizes phase and interference problems by locating the sound source (all 3 drivers) close to the ideal ear level. The aperiodic feature is obtained by using a highly damped vent (not a reflex port) which is carefully controlled by filling a narrow slot with a precise density of fiber glass. The acoustical resistance lowers the Q of the system through this high friction venting; and, as a result, a more linear load is presented to the amplifier. This ensures more acoustic output from your amplifier than would be possible with other systems of the same nominal efficiency but with typically larger impedance variations. The system uses two 10" woofers and a dome tweeter which provides slightly smoother midrange, better bass, and superior capability at greater power levels. Specifications are: rated impedance 8 ohms; crossover frequency 1,000 Hz; 5 high frequency level positions; rms power rating 75 watts; dimensions, 28" H, 21½" W, 10" D. Refer to Fig. 7-1 for illustration.

Grommes model GS-310, full-range 3-speaker system is carefully engineered to provide natural sound through a low resonance air suspension 10" low frequency, special 6" midrange, and wide dispersion 3½" high frequency speaker. Variable mid and high frequency controls allow precise adjustment of presence and brilliance for any room acoustic environment. Response is smooth from 25 through 20,000 Hz and 3-way network provides crossover frequencies at 600 and 3,500 Hz. Power handling capacity is 50 watts program material. Dimensions are 23" H, 13¾" W, 10½" D. Finish is hand-rubbed walnut and the unit may be used on floor or shelf, either vertically or horizontally.

Fig. 7-1. Dynaco A-50 aperiodic speaker system. (Courtesy Dynaco Inc.)

Fig. 7-2. KLH model 5 speaker system. (Courtesy KLH Research & Development.)

KLH model Five, compact speaker system is a 3-way, 8-ohm, with a 12-inch driver operating up to 600 Hz. Here the two 3" cone speakers take over up to 3,000 Hz, followed by the tiny 1¾" direct-radiator tweeter handling frequencies above 3000. A pair of switches on the back of the cabinet allow 2 to 3 db change in the level of midrange or high-frequency speakers to accommodate the listener's conditions. The smoothness of coverage may be noted by the plus or minus 2.5 db from 170 through 14,000 Hz which is actually better than a high-quality phono cartridge can deliver. The acoustic suspension woofer shows smooth, strong performance down to 20 Hz with only 6 percent distortion at that level. Dimensions: 26" H, 13¾" W, 11½" D and although deep sturdy shelf installation is possible, weight and size suggest floor mounting. See illustration of Fig. 7-2.

Fisher model XP-44B 2-way speaker system uses a 6-inch woofer and 3-inch tweeter. Electrically matched to provide perfect performance with modern day solid-state amplifiers, it offers flat response from 39 through 19,000 Hz. The unit is enclosed in a walnut vinyl cabinet measuring 8¼" H, 15¼" W, 6¼" D and weighs only 15 pounds.

Fisher model XP-9C, 4-way, 5-speaker full-size system employs 15-inch woofer, two 5-inch midrange, 1½ inch dome tweeter, and 1½ inch dome super tweeter. Frequency response is 28 through 22,000 Hz and a 3-position treble acoustic balance control is provided. The walnut cabinet measures 16¼" H, 27½" W, 13" D and the unit weighs 55 pounds.

Harman-Kardon model HK25 omnidirectional speaker system features high performance in a 2-way with LC crossover network and brilliance level control. Easily placed where they look and sound best with 360-degree sound assured; floor, wall, or ceiling suspension may be selected. The acoustic suspension design enables greater power to be comfortably handled without distortion even while deep bass is being dispersed. Rigid polystyrene foam construction and simulated walnut vinyl finish produces an attractive, sturdy unit. Frequency response ranges from 40 to 20,000 Hz with a power capacity of 35 watts. The 6-inch woofer and 2¼" tweeter offer an impedance of 8-ohm. Enclosure is 16½" high with a diameter of 12¼".

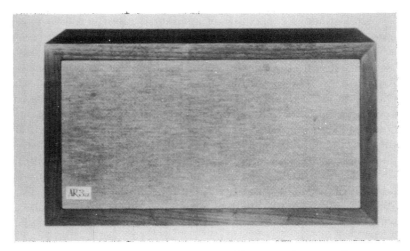

Fig. 7-3. AR-3a 3-way speaker system. (Courtesy Acoustic Research, Inc.)

Acoustic Research model AR-3a, 3-way speaker system was designed to reproduce music as accurately as present-day knowledge of acoustics and electronics permit. The 12" bass driver, 1½" dome midrange, and ¾" dome tweeter allow the laws of physics to operate to the listener's advantage while spreading frequencies through a wide angle. This affords considerably more realistic reproduction of music to listeners in all parts of the room. The frequency response is extremely smooth over the entire range from 30 to 17,000 Hz with recommended amplifier power of 25 watts rms per channel. The impedance is 4 ohms and crossover frequencies are 575 Hz and 5,000 Hz. Cabinet measures 14" H, 25" W, 11½" D and weighs 53 pounds. The system is priced at $225.00 in unfinished pine or slightly higher in mahogany, birch, oiled walnut, glossy walnut, oiled teak, or cherry. See Fig. 7-3.

Toshiba model SS-36, 4-way speaker system with 12" air-suspension woofer having a free-edge cone produces superb bass at every point from 30 through 1,000 Hz. The 6½" midrange covers 1,000 through 5,000 Hz, the 3" by 1½" horn-type tweeter covers the lower treble at 5,000 to 9,000 Hz, and a 1½" super-tweeter takes over at 9,000 through 20,000 Hz. Specifications are: impedance 8 ohms; frequency response 30 to 20,000 Hz; power handling capacity 30 watts (60 watts peak); controls: midrange level, tweeter level; multi-channel terminals. 2-way and 3-way; dimensions, 25⅝" H, 15⅜" W,

11⅞" D; weight 33 lbs; walnut enclosure (acoustic suspension).

KLH model twenty-three, 2-way acoustic suspension system establishes a new standard of performance and value in the medium price range. Extraordinary transient response and high frequency dispersion are apparent. The 12" low frequency speaker has a heavy magnet assembly and the high frequency is a 1¾" direct radiator with a cone of critically shaped shell-like material which spreads the coverage to the midrange and eliminates need for a separate driver. Specifications include: impedance 8 ohms, 25 watts IHF music power (min.), dimensions 25¼" H, 14½" W, 11½" D. A 3-position switch permits 2.5 db change in level of high. See Fig. 7-4.

Fairfax model FX-100A is a two speaker 2-way system with ducted port design having a heavy duty 8" wide-range driver with excellent linear movement. Using a special 3" tweeter, the system offers a frequency response of 32-20,000 Hz and has a 30-watt input with an 8-ohm impedance. Minimum power requirement is 8 watts: dimensions are 21" H, 12" W,

Fig. 7-4. KLH model 23 2-way acoustic suspension. (Courtesy KLH Research & Development.)

Fig. 7-5. Fairfax model FX-100A 2-way speaker. (Courtesy Fairfax Industries, Inc.)

7⅞" D. Weight is 22 lbs. It is priced at $79.95 and is shown in Fig. 7-5.

Altec model 887A acoustic suspension speaker is designed to offer a wide frequency response with low distortion and precise musical balance along with small size. The 2-way Capri uses an 8" low frequency speaker and a 3" direct radiator for high frequency reproduction. It is shown in Fig. 7-6. Size is 19" W, 10" H, 9" D.

REFLECTOR SPEAKER SYSTEM

The Linear Design Labs model 749 with its electro-mechanical design concept produces an exacting reproduction of natural sound. Looking under the rear grill cloth, you will find, not four—but eight efficient, high-compliance, 4-inch speakers. In the front, a single speaker of the same design and size, using no woofers, tweeters, equalizers, or crossovers. The eight speakers at the back are mounted in clusters of four with each angled precisely on a sound dampened panel. Since sound impulses are routed directly to each speaker, it is unnecessary to employ the adjuncts so commonly found in other multi-speaker arrangements. The mounting angle of each individual speaker results from countless hours of research plus unyielding test determinations under varying acoustical conditions to achieve these optimum sound-expressions. Secured by the fact that 89 percent of the sound output is projected against the walls and reflected into the listening area, the astonishing effect of reflected sound found only in concert halls can hardly be a surprise.

AIR SUSPENSION SPEAKER SYSTEM

The Pioneer model CS-44 represents a highly efficient design of the air suspension type, and is capable of being driven with inputs as great as 25 watts with a sensitivity of 96 db per watt to make it ideal for either low or high-power amplifier applications. It may be used as the main speaker in a hi-fi system, and also adapts well to tape recorder or auxiliary speaker operation. The woofer is designed to operate without cone breakup at extremely high audio input power. The soft surround ensures optimum compliance, and as the 8" woofer has a long-throw voice coil, the power handling in the bass register is clean, with no muffle or break-up. Considering its size, the bass is unusually responsive. The high-frequency tweeter is 2½" and a newly developed cone type with wide-angle dispersion. Response to frequencies in the 10 to 20 kHz

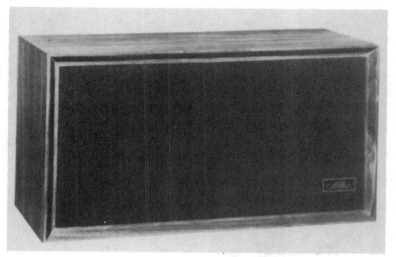

Fig. 7-6. Altec model 887A Capri acoustic speaker. (Courtesy Altec Lansing.)

Fig. 7-7A. Utah 3-way speaker system (inside view).

Fig. 7-7B. Utah model HSI-C speaker system. (Courtesy Utah Electronics.)

region ensures reproduction of all important harmonics and transients, to provide a major plus to the fidelity and quality of sound. Peaking at 25 watts but still exceptionally efficient with lower driving power, this speaker handles transient bursts even in the bass region with incomparable smoothness of reproduction. The decorator inspired cabinet is meticulously constructed of rich veneers in oiled walnut. Specifications include: impedance 8 ohms; crossover frequency 2,500 Hz; frequency response 35 Hz to 20,000 Hz; maximum power input 25 watts; sensitivity 96 db per watt; dimensions 11⅛" W, 9¾" D, 19⅛" H; weight 18 lbs.

The Harman Kardon model KH10 is a bookshelf, air suspension system designed to deliver maximum bass response with very low distortion. The driver utilizes an extended range cone that projects smoothly to beyond 15,000 Hz. This small speaker has an unusually high power handling capacity, and manages 20 watts of music power without breakup. Response is virtually flat from 40 to 18,000 Hz. Impedance is 8 ohms. Dimensions are 9" W, 14" H; 7½" D for its hand-rubbed oiled walnut enclosure.

An inside view of the Utah 3-way speaker system is shown in Fig. 7-7A along with the outside view of the HSI-C model in Fig. 7-7B. The 12" woofer has a 1½" voice coil and the system

Fig. 7-8. Utah model A-130 3-way speaker system. (Courtesy Utah Electronics.)

165

Fig. 7-9. KLH model 17 speaker system. (Courtesy KLH Research & Development.)

offers a frequency response of 30-19,000 Hz. It is priced at $99.95.

The Utah model A-130 is a 12" 3-way acoustic suspension system that provides the latest in enhanced listening environment. Separate controls for midrange and tweeter permit system response to adjust to your musical taste. It is priced at $129.95 and is pictured in Fig. 7-8.

The KLH model 17 is a compact speaker system that delivers excellent performance which is made even more remarkable when size and cost are considered. Using a 2-way acoustic suspension system with a 10-inch woofer and a completely sealed, internally padded enclosure $11\frac{3}{4}$" by $23\frac{1}{8}$" by 9", it is priced at $74.95. The model 17 is pictured in Fig. 7-9.

POWER HANDLING CAPABILITY

The power actually applied to your speaker system must never exceed that specified by the manufacturer in terms of comparable ratings. These figures will vary according to the standards and-or conditions under which they are formulated.

Music power may be rated by IHF standards based on power output with a maximum of 1 percent total harmonic distortion (THD), or EIA ratings based on THD as high as 5 percent. Power ratings are always expressed in watts (W), but rms figures will be about half the peak figures for the same amount of power, so make sure of this when comparing ratings. In some cases, both rms and peak values will be given, but if only one is given without specifying, be sure to ask. Now, if you have rms for one unit and peak for the other, you may multiply the peak by .707 to change it to rms, or rms by 1.414 to change it to peak. The stereo amplifier output power may be the total for all channels unless the condition is specified such as 15-15 which would mean 15 watts for each channel of the 2-channel amplifier and the 4-channel amplifier could indicate 15 watts for each channel by 15-15-15-15. Total IHF power 44 watts could mean 15-15 rms power, as the 44 watts indicated total (peak) IHF power must be multiplied by .707 to change to rms power, followed by dividing by 2 for the individual channel rating. The impedance of the amplifier output is always considered when giving output figures, and the speakers used must match that rating to be significant. If the amplifier delivers 15 watts per channel into 8 ohms, it will deliver much more into 4 ohms and much less into 16 ohms, so find out the exact amplifier output impedance for your speakers. Most speaker systems have an impedance of 8 ohms, although some may be rated at 4 ohms.

The manufacturer of the speaker determines its electrical power handling ability by the temperature rise of the voice coil when a voltage in the form of a sine wave is applied. Power is increased until an established safe temperature is reached at which point the power is calculated. If that power is measured with an rms meter, the rating will be in rms watts and a peak rating would be double.

ALLOWING A SAFETY MARGIN

You may not realize that a speaker rated at 20 watts for example, may not be capable of handling that amount of power at every frequency within its response limits, but can operate at full rated power during a typical or average musical rendition. However, if a speaker is able to handle 20 watts of power at a specified frequency, chances are that it may do considerably better on an integrated program. It should be pointed out that speaker ratings in watts can be quite misleading, and such figures, unlike electrical power measurements, are not firm, and will vary according to the frequency of the applied signal, and the opposition to current flow at that frequency. So 20 watts at 1,000 Hz would be less than 3 watts at 100 Hz. Since we deal in integrated program material, and music power terms, a rating expressed in peak music power would be halved when compared to rms music power. Safety limits are pegged by the temperature of the voice coil when continuous power is applied and as power is increased, the temperature of the coil rises until a safe limit is reached. Measuring power at that point establishes a reasonably safe limit as music signals from your amplifier do not present continuous peaks, and there will be considerable cooling time between the peaks in practice. This ensures a satisfactory margin for safety, and in an airtight enclosure an even higher level of power is permissible without damage as cone travel is reduced. Connecting a VTVM across the enclosed speaker while advancing power until distortion begins will provide a peak voltage reading that may be converted to watts to indicate a safe margin under these conditions as rapid peaks were not shown on the meter.

POWER AND ITS EFFECT ON WHAT YOU HEAR

Space limitations, such as those imposed by a small room, make it difficult to achieve the high quality of reproduction that is to be expected from a worthwhile stereo system. This is due, in part, to the inefficiency of the shelf type speaker system which it is necessary to use, and as a direct result, much more power is needed to overcome this deficiency. So it becomes obvious that where space is at a premium, some sacrifice must be anticipated in the effectiveness of the arrangement. Many of the better speaker systems offered for bookshelf or small room coverage, have a lower impedance which requires additional power from the amplifier to compensate for the losses in such an environment. Furnishings in the small room category will accentuate these losses and make the delivery of more power from the stereo system an essential factor to achieve adequate sound reproduction and quality. Actually small speakers perform best in small rooms where they can handle the surrounding area just as the large speaker systems are best suited to large rooms where they can be physically accommodated and used to best advantage. If a mismatch exists between the speaker and the amplifier output, even a small impedance mismatch can result in some loss in power although the problem is less serious if the speaker impedance is higher than the amplifier impedance. This will not result in distortion, but rather a reduction in output power available from the amplifier to a degree according to the actual amount of difference involved. In the opposite direction where the speaker impedance is lower than the amplifier impedance, distortion and overloading of the amplifier may become quite serious.

GETTING MORE "LOWS" FOR LESS MONEY

Probably the simplest way of improving the low end of your speaker response rests in the proper placement in your particular listening area. Actually this amounts to a trial and error procedure and can best be perfected by the gradual movement of the speaker closer to the wall and adjacent corner of the room or shelf. Moving the unit a few inches at a time while listening to music with plenty of low passages, should quickly pinpoint the exact spot for best low frequency response of your particular speaker system. If additional bass is still desirable without extra expenditures, you may try reversing the two wires or leads to the speaker in question even though you feel that phasing is correct. The possibility exists that filters or crossover networks have upset the phasing pattern, but in any case, it is quite simple to change back if no improvement is noted. There are many other ways of emphasizing bass response but most of the gimmick arrangements will reduce the fidelity of your overall system to an unacceptable degree. Assuming that your amplifiers have the capability to handle more bass material, then the speaker system or at least part of it should be changed to a unit that is better able to underline that portion of the audible spectrum in a natural, rich-sounding manner. The average musical renditions seldom carry below 50 Hz which is the rapid drop off point for many speaker systems, and even the bass tuba cannot get as low as 40 Hz.

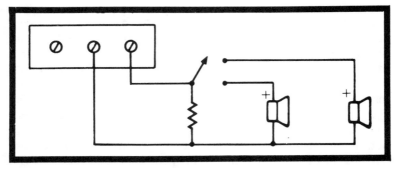

Fig. 7-10. Speaker switch for solid-state amplifier.

PROPER USE OF EXTERNAL CONTROLS

Special selector switches are necessary for solid-state amplifier outputs and these may be easily connected at convenient locations as shown in the sketch of Fig. 7-10. This allows selective switching between any one of three pairs of speaker systems, and through the use of a top quality, non-shorting type of contact, these switching units are quite capable of handling considerable power without danger of malfunction or damage to the stereo components. An extremely handy and attractive external balance control utilizing dual concentric L-pads and two phone jacks mounted on a brass plate while providing screw terminals in the rear offers a simple way to connect and install the unit. It may prove quite useful for controlling volume on a second set of speakers which may be plugged in from the front without upsetting system balance. The extension speakers may be connected or removed easily at your convenience by plug-in arrangement without affecting the main part of the stereo system. Although normally available in 8-ohm values, others may be found in most stores or simple modifications made according to instructions provided. In extending speakers and controls, proper connections are usually very easily made and information is normally supplied with the units involved, but suggestions as to the wire used for such convenience controls and additional extension speakers may save loss of power as well as eliminate other problems that could result. Speaker wire as usually offered in hi-fi or TV shops is actually a miniature lamp cord parallel wire that is satisfactory for speaker use for power of only a few watts and for short distances. Never use it for electrical power, or for distances greater than 15 feet without line transformers. This small speaker cable comes in gray, white, brown, and clear with 2 wires (No. 24 guage, stranded) and is very flexible. But the loss due to the small gauge can become terrific according to the power applied and the actual distances involved. Regular lamp cord (zip cord) has two No. 18 stranded conductors and although far less flexible than the smaller speaker wire, is ideal for longer speaker system extensions except where extremely high amplifier outputs may be required and speakers of lower impedance (less than 8 ohms) are used. This wire is normally found in brown, black, gray, or white and also in larger sizes with the No. 16 gauge available at most stores. As an indication of the possible loss in power by using the small wire (No. 24) in place of the large (No. 16) for an extension speaker in an adjacent room, you could lose 90 percent of power, and that is ignoring heat loss. So make sure that the wire size is proper for the distance and power as well as the impedance of the speaker system. Remember that impedance, often indicated by the letter Z, is the opposition to current flow of all types except direct current (DC). Measured in ohms, it applies to audio which is merely a form of alternating current.

IMPORTANCE OF FREQUENCY RANGE

When a fast drop-off in frequency response occurs in a speaker system, the deficiency in the audible sound becomes obvious to the average ear. A tinny sound results if the low end of the range cuts off early; or a muffled, overly suppressed sound when premature high frequency fall-away exists. Even the mid-range excessively boosted or accentuated can result in an annoying repetitive effect with lingering extensions of sound, all of which make the wider range coverage by the speakers quite important to the average listener. An occasional dip of short duration does not offend, and in fact may not even be noticed, but the longer duration irregularities resulting from sharp crossover attenuation, may be most disturbing. This emphasizes the desirability of an extended pass band to allow some overlapping of the two separate drivers involved in the speaker response. Realizing the variations in opposition (impedance) offered by the speaker system according to frequency, and the fact that this part of the hi-fi system is much slower acting than any other segment or combination of segments, it can readily be understood that perfection here, though most desirable, is exceptionally difficult if not impossible to achieve. So we compromise, settling for a unit that comes as close as possible without being over the limitations of the budget. In this regard it may be advisable to remind anyone looking for a speaker system to improve their overall stereo to avoid selecting a $250 speaker to connect to a $150 system as you simply would not have the quality of output from your amplifiers to do the speaker justice. It couldn't possibly offer you more quality out than you were putting in, so when ugrading parts of your system, always work from front to back. In other words, improve the program source first (AM-FM tuner, record player, or tape player), then preamplifier or amplifier, and finally the speaker system. If you are able to improve the complete system by replacing all parts at once, or if starting with your initial system then check each unit carefully by looking over the manufacturer's descriptive literature and specification sheets as well as firing questions at the salesman to make sure that the system is compatible. You may also want to consider possible future developments in hi-fi and whether the system under consideration could be readily adapted to these potential changes. Naturally some attention must be paid to 4-channel which is definitely here to stay, and even though you may not be ready for it at this time, do not close the door to some form of adaptation in the future. Advances in design and techniques have always met with stern resistance from the consumer for numerous reasons, as was quite evident in the case of the vacuum tube and its replacement with solid-state devices. Although it soon became quite obvious that reliability was a major advantage to be gained by the use of transistors and diodes, the old-fashioned tube approach persisted to be demanded by the consumer because it was just as good in the beginning and about 20 percent cheaper. A couple years later, the tube-device had been to the shop for repairs at least once, and the solid-state device in most cases had never required repair and showed no signs of needing such services (expense) in the future. This is not intended to evaluate 4-channel sound as a money-saving technique, but merely to emphasize the importance of preparation for possible adapters or converters being easily added to any prospective system as may be needed in the future to cope with developments.

ADDITIONAL SPEAKER SYSTEMS

The Acoustic Research model AR-LST is a speaker system utilizing nine drivers and provides an arrangement for the audio professional as well as the perfectionistic music lover. This 3-way system with 9 speakers includes an acoustic suspension 12-inch woofer, four 1½" hemispherical-dome midrange, and four ¾" hemispherical-dome tweeters. The impedance of the system according to the setting of the six-position switch, which varies the filter networks and their respective speakers, may vary from 4 ohms to about 16 ohms. This switch adjusts the relative level of highs plus varies

Fig. 7-11. AR-LST driver arrangement. (Courtesy Acoustic Research, Inc.)

Fig. 7-12. AR-LST speaker system. (Courtesy Acoustic Research Inc.)

system impedance by permitting an auto-transformer to add impedance at middle frequencies. The speaker arrangement is shown in Fig. 7-11 with the 12-inch woofer in the center panel, the four tweeters across the top of all three panels, and the four midrange drivers below in either side panel. The actual dimensions of the speaker system are $27\frac{1}{8}$" wide, 20 inches high, and $9\frac{7}{8}$ inches deep. The speaker can handle peaks exceeding 500 watts without distortion and are protected against any possible power damage with fuses. The system is priced at $600. The AR-LST is pictured in Fig. 7-12. The simple crossover network connection to keep the high audio frequencies out of the woofer and the low audio frequencies out of the tweeter is illustrated in the sketch of Fig. 7-13. The coil, or inductance as it is commonly referred to, acts in the opposite way to the capacitor; as the frequency is increased, the impedance offered by the coil increases while that of the capacitor decreases. Naturally this is exactly the action required in the conventional crossover network, as the speakers would otherwise all be subjected to the same frequencies. The result would be distortion and probable damage without this action. The coil has little or no effect on the low frequencies passing through to the woofer but offers such opposition to the high frequencies that they take the low resistance path through the capacitor to the tweeter. The low frequencies meet so very much opposition at the capacitor that they take the path of the least resistance to the woofer.

The Sansui AS-200 acoustic suspension speaker system is the result of extensive research for a means of more fully exploiting this type of design. It features extended bass response with a rich well-damped bass sound and natural tonal quality along with the wide dispersion of reproduced sound. The acoustic suspension design makes a much more compact cabinet possible to provide considerable flexibility in positioning the speaker in the room. This 3-way 3-speaker system has a 50-watt power capability and the airtight enclosure houses a high-compliance woofer, a large cone midrange, and the new-type cone tweeter. The crossover network is an LC type that produces a 6 db-combination cutoff characteristic. Keeping undesirable phase rotation of reproduced sound to a minimum, it enables the three speakers

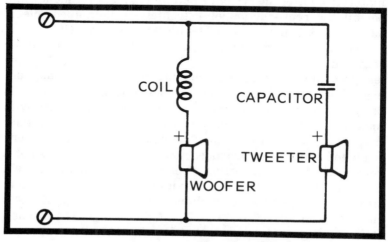

Fig. 7-13. Simple woofer-tweeter crossover circuit.

Fig. 7-14. Sansui model AS-200 speaker system. (Courtesy Sansui Electronics Corp.)

to reproduce the entire range of sound with natural transitions. Continuously adjustable volume-type attenuators are provided for the midrange and tweeter to allow control of the tonal quality according to the listener's preference or room acoustics. The enclosure utilizes special sound absorbent material to lower the system's bottom resonant frequency and permit the reproduction of extremely well-damped, clear, deep lows. Side panels are of 19mm-thick high-density, superhard chipboard with 22mm-thick rear and baffle panels of the same quality material, mitre-joined to ensure that they are leakproof and then further reinforced to preclude any undesirable resonance. The large cone midrange speaker is designed to reproduce the 2,000 to 7,000 Hz range with maximum efficiency, cutting out the lower and higher frequency ranges where intermodulation distortion is very likely to happen. This produces powerful, rich midrange notes completely free of distortion. Specifications are: woofer 10", midrange 6½", tweeter 3"; maximum power 50 watts; sensitivity 86 db; impedance 8 ohms; frequency range 40-20,000 Hz; dimensions 12⅜" W, 23½" H, 11⅞" D; weight 34.1 lbs. net. Refer to Fig. 7-14 for outside view of the AS-200 speaker system.

The University Sound Division of Altec, model Concept EQ-5 consists of two 3-way bookshelf speakers and an active equalizer in a walnut cabinet. Designed for amplifiers with 40 continuous rms watts per channel or more and featuring unusually high power-handling capability to let the amplifier release its full potential. The system overdamps the bass driver to allow a natural roll off of bass frequencies below 150 Hz, allowing the air in the enclosure to control speaker movement down to 30 Hz. As the speaker follows its normal roll off characteristic, the Concept equalizer applies a reverse frequency response curve to the electrical signal prior to amplification. The net result is system reproduction of frequencies below 50 Hz with harmonic distortion of less than one-tenth that of conventional designs and excellent transient response. Bookshelf design of the speaker cabinet permits

Fig. 7-15. University Sound Concept EQ-5 system. (Courtesy University Sound Div. of Altec.)

Fig. 7-16. KLH model 33 2-way speaker system. (Courtesy KLH Research & Development.)

complete freedom of placement as the equalizer has controls for proper balance of bass response with wall-shelf, floor, or corner arrangements. Further flexibility is achieved with a unique mid-bass control which allows the user to change the system contour to adjust for the room gain or acoustics of the individual listening room. Adjustable high frequency control is provided for variance in room liveness or brightness. Specifications include: frequency response plus or minus 2 db from 30-20,000 Hz with equalizer; power rating 100 watts; impedance 8 ohms; dimensions 25⅛" W, 15⅛" H, 10⅝" D; weight 30 lbs. each unit. It is priced at $399 complete, the Concept EQ-5 is pictured in Fig. 7-15.

In the $100 class, the KLH model 33 2-way loudspeaker system reproduces the entire audible range with such low distortion and musical balance that its performance passes all expectations for a speaker of this size and cost. It is housed in an oiled walnut cabinet of elegant simplicity which measures 23⅜"H, 12¾"W, 10 5/16"D and finished on four sides for either horizontal or vertical placement. Grille panel is removable for cleaning or substitution of other grille material. The controlled acoustic compliance permits optimum impedance matching of the woofer to the amplifier in the 40 to 120 Hz region for maximum power transfer and precise control of the woofer cone excursions to ensure minimal bass harmonic distortion. This feature makes possible selective stiffening of the air enclosed by the cabinet by increasing air molecule motion in the range between 40 and 120 Hz. The increase in stiffness is carefully controlled by two factors: (1) reduction of the Q by supplying a resistance to the movement of air molecules within the enclosure; (2) a resistive damping grille used to regulate the movement of the air mass in a specially designed, broadly tuned cavity. The result is a woofer so linear in performance that its low bass distortion is unprecedented for a woofer of its size and cost. The three-position switch on the crossover permits adjustment of high frequency response to suit almost all program material and listening environments. Priced at $99.95, the model 33 is pictured in Fig. 7-16.

Altec's model 891 bookshelf speaker system is a compact, high-quality, 2-way system designed to meet the highest standards of performance. This new dynamic force speaker delivers a high level of performance with superior low

Fig. 7-17. Altec model 891 bookshelf speaker system. (Courtesy Altec Lansing.)

Fig. 7-18. Utah model A70-A 3-way speaker system. (Courtesy Utah Electronics.)

frequency resolvement and low distortion characteristics. Components include a 12-inch high-compliance dynamic force woofer, an Altec exclusive, using a foam surround and copper voice coil. This woofer is designed specifically for operation in a sealed, infinite baffle cabinet to provide extended low frequency reproduction. The high frequency transducer is a 3-inch direct radiator with small diaphragm area for best wavelength resolvement, extended smooth response, and excellent wide angle dispersion. Maintaining medium high efficiency, only 12 watts are required for full, roomfilling sound and capacity is such that much larger amplifiers may be used without causing speaker damage. It is finished in handsome hand-rubbed oiled walnut and priced at $125. The pictorial view is displayed in Fig. 7-17.

Utah model A70-A is a 10-inch 3-way acoustical suspension speaker system offering high performance combined with decorator styling. Ideal for bookshelf duty but large enough to be placed on the floor, the 10" bass speaker incorporates a 2 lb. ceramic magnet structure. A 6" direct radiator midrange and 3½" tweeter complete a system which offers a distinctive and dramatic sound at a surprisingly low price. Cabinet is genuine walnut veneer on ¾" stock; size 12½" H, 22" W, 9" D. Specifications are: frequency response 38-18,000 Hz; power handling 25 watts (50 watts peak); impedance 8 ohms; and priced at $79.95. The model is pictured in Fig. 7-18.

The KLH model 32 is a 2-way acoustic suspension speaker system that reproduces the audible frequency range with such low distortion and outstanding musical balance, even at high playback levels, to be unprecedented in a speaker at this price. The electro-acoustic efficiency will permit you to exploit the full capabilities of current low-priced receivers in the 10-watt steady-state per channel power class. However, the speaker will handle short duration peaks up to 100 watts in high quality systems, and when used as a remote or extension speaker, its volume level will closely approximate that of the most modern acoustic suspension systems. The woofer has a four layer voice coil wound on an aluminum former in a narrow-gapped, large magnetic structure for optimum efficiency. The tweeter is a 2-inch direct radiator with dome

Fig. 7-19. KLH model 32 two-way speaker system. (Courtesy KLH Research & Development.)

providing exceptionally level and flat response, free from ringing or other spurious outputs, to well beyond audibility. Dimensions are: 19⅜" H, 10⅞" W, 7¼" D. It is priced at $47.50 each. The model 32 is pictured in Fig. 7-19.

The speaker system is the final link in the long chain of equipment and events leading from original performance to the reproduction of music for the home. The ultimate result will depend on the capability of each link in the chain to accurately interpret and faithfully respond from input to output, including this final link referred to as the speaker system.

Chapter 8
Speaker Enclosures

The trend in the past was to hide speakers behind drapes, furniture or other articles in the room, but experience has indicated that this approach was far from ideal. As speakers are now constructed in a more attractive design along with a functional character, the room actually takes on a rich appearance when the speaker is placed out in full view, and the sound is in the prime listening area as the very musicians creating it. No muffling, no lack of clarity, only true reproduction as the avid listener enjoys it most. The best placement for your speakers is frequently dictated by the directional properties as well as the end result desired by the listener, as in the case of the omnidirectional unit. Here we require open space all around, and the proximity of walls or corners is bound to restrain the free flow of sound waves so essential to the design features intended. Any restrictions resulting from drapes or other sound-absorbing objects should always be avoided.

Manufacturer's specifications will normally suggest the proper or intended location for any of his units in order to realize optimum performance, and by checking into these suggestions when making your selections, the speakers for the desired purpose may readily be chosen. Just as some speakers may not be mounted on bookshelves, others may not be placed on the floor or table, and wall mounting could be out of the question for certain types. By carefully considering the exact spot in your living room where most of the relaxing will be done, adequate coverage from your stereo system may be judiciously planned. Proper speaker arrangement will afford excellent four-channel sound for any specific listening area in your home; plan your layout and follow that plan to the letter with at least the minimum equipment necessary to do the job.

WALL, CLOSET, AND SHELF ENCLOSURES

Bookshelf speaker enclosures may be constructed from ¾" thick bookshelf wood which normally has a width of about 11 inches. This stock is solid with a smooth surface and may be easily stained or covered with vinyl or cloth following assembly. The finished stock may be even more desirable as it is similar in appearance to the wood paneling of some station wagons and has the same thickness as the unfinished stock but may vary in width from 6" to 12". Plywood has many objectionable characteristics and is not recommended for speaker enclosures. The solid stock is easy to cut and inexpensive for the purpose, requiring only a few cuts with the saw to supply enough panels for the cabinet.

Hi-fi acoustic suspension speakers of the 8" inexpensive variety may be used for many bookshelf of closet type enclosures and could also serve for wall mountings with sufficient baffle. These speakers often run as low as $8 and have a power handling capacity of about 30 watts at 8 ohms. Frequency response is satisfactory from 30 Hz to 20 kHz. The larger 12" coaxial speakers are available from $15 up and have a frequency response of 23 Hz to 19 kHz with a power handling capacity of 45 watts at 8 ohms. These would be ideal for closet installations and could also prove worthwhile for some wall type layouts.

The assembly of an enclosure for a 2-way speaker system should not be difficult for many who prefer to build their own speaker systems. This is one of the better ways to get things exactly as you want them and at the same time to save a few dollars without sacrificing quality. Two important requirements of any speaker enclosure are the rigidity and air-tight construction. Ordinary bookshelf wood ¾" thick will provide the desired stiffness, and although the back panel may have a tendency to vibrate, it may be stiffened sufficiently by bracing with a couple pieces of lumber screwed to the inside. Since the edge around the front panel may leak air due to the grille cloth in the corners, a little caulking compound or even putty pressed here from the inside should take care of this problem nicely. Mitered corners are desirable, but not always possible with limited tools, so a screwed butt joint plus a little glue could suffice. If rails are used front and back, using ¾" wood, the smallest outside dimension in front should be 3 inches larger than the speaker diameter which allows ¼" space between the edge of the speaker and a ½" rail.

The layout of the front panel should be as shown in Fig. 8-1 with the woofer centered in the bottom portion and the tweeter as high as practical. The port opening should be positioned to yield maximum strength by allowing ample wood between port and woofer cutouts. Sound deadening is best accomplished by using one-inch thick fiberglass or a similar porous material on all inside surfaces except the front panel. The fiberglass does not need to be closely fitted, a 2" border is close enough and this material is ignored in figuring the inside cabinet volume. In order to avoid grille cloth buzzing, the cloth should be glued on the front panel with Elmer's glue which does not dry before you get a chance to stretch the fabric and staple it in place.

When determining the length of tubing required to bass reflex your cabinet you must know the free air resonant frequency of your speaker as well as the internal volume of the cabinet not counting the fiberglass. The outside dimensions of the cabinet suggested are shown in Fig. 8-2 and after subtracting the thickness of the wood, it is 14" H by 11" D by 22½"

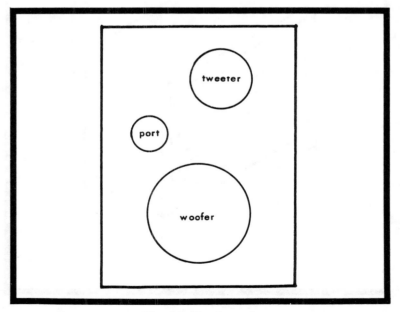

Fig. 8-1. Front panel.

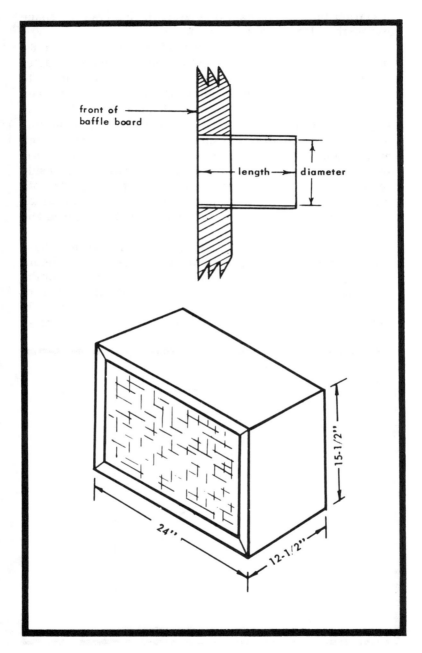

Fig. 8-2. Cabinet.

W equals 3,465 cubic inches. Dividing by 1,728 to change cubic inches to cubic feet, shows about 2 cubic feet. If alternatives are available, the mailing tube diameter which corresponds to a length approximating the speaker radius should be selected.

It must be remembered that the tweeter is installed near the top of the panel arrangement for best performance at the frequency range, but have a crossover installed to avoid permanent damage. Due to the high frequencies handled, cone travel is limited by very stiff suspension and when a low frequency is applied it tries to move the cone a long distance. The tight suspension arrangement prevents such action and the voice coil acts like a frozen electric motor, it simply gets hot and burns out. You may recall that high frequency power represents the minor portion of total music power, so the small tweeter is adequate with a heavy duty woofer—if a series capacitor at least is connected in one side of the line feeding the tweeter to keep those lows out of the voice coil. A nonpolarized (NP) electrolytic capacitor having a value of 4 mfd and connected as shown in Fig. 8-3 with an 8-ohm tweeter will provide a crossover frequency of 5,000 Hz which is where the impedance of the capacitor is equal to that of the speaker.

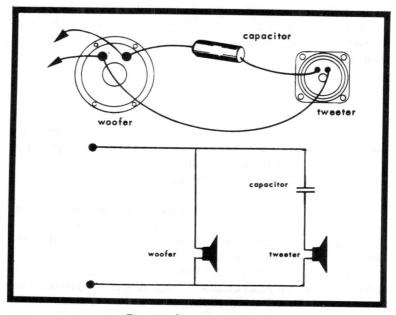

Fig. 8-3. Simple crossover.

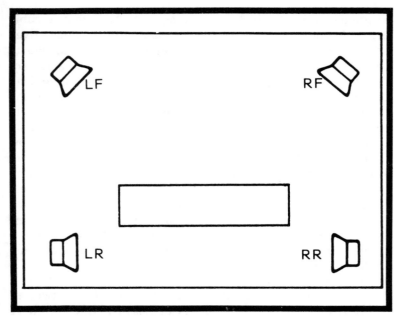

Fig. 8-4. Four-channel speaker arrangement.

At this point the tweeter output is down 6 db and will be further reduced by 6 db each time frequency is halved, which indicates the easy way of protecting the tweeter against damaging lows.

ASSEMBLING SPEAKER CABINETS FOR UNUSUAL BALANCE

The 8-inch 2-way acoustic suspension speaker systems are large enough to handle the left and right rear assignments, and yet small enough to fit unobtrusively into their respective corners at the back of the room, facing each other as shown in the layout of Fig. 8-4. In order to complete the 4-channel system, the 10-inch 2-way acoustic suspension speaker systems carry the left front and right front channels from their place in the front of the room as also indicated in the sketch with each unit facing the listening area.

The HK50 omnidirectional speaker systems used as end tables do offer unusual balance in a fine stereo system. They have a power handling capability of 40 watts with a frequency response of 35 to 20,000 Hz. Sound is dispersed evenly throughout the listening area by the reflector which faces the heavy duty 8" woofer and wide dispersion tweeter. All connections are made on the bottom, near the crossover network and high frequency level control. The hand-rubbed oiled walnut enclosure surrounds the solidly constructed acoustic suspension cabinet. This arrangement promotes a feeling of increased depth as a result of the reflection pattern of middle frequency tones.

SUSPENSION ARRANGEMENTS

Just as sound separation affords the real test of any speaker system, the acoustic suspension woofer and tweeter provide fantastic off-axis performance. The 2-way system utilizing this principle offers clean, realistic bass response and clear, crisp highs as well. In the 3-way arrangement, midrange is covered by a direct radiator, highs by a super-tweeter, and the natural bass through a carefully designed woofer. Thus reasonably flat response may be expected over the entire frequency range. The acoustic suspension principle for the woofer extends bass response at low distortion by replacing the mechanical stiffness of the air trapped within a completely sealed chamber. This air spring has proven to exercise exceptional linearity in control of cone motion in the bass region, showing superiority over the best conventional suspensions. It actually reinforces the ability of the cone to display accurate movement over the long distances required to push air at very low frequencies. However, smoothness of response is necessary over the entire range of the low frequency speaker and roll-off at the high end must be precisely controlled. This dictates careful design of cones and suspensions as well as other critical parts of the unit. Such quality control and uniformity makes it possible to easily match pairs for optimum system performance.

The high frequency speaker, in order to meet requirements of precise motion and adequate power output at high frequencies as well as wide dispersion of sound at all frequencies, utilizes a cone that is light and small in diameter. When attached to the outer rim by liquid rubber, the cone may move accurately up to one sixteenth of an inch, and the exceptional freedom of movement increases power handling capability over a much wider range of frequencies. Thus the

Fig. 8-5. 2-way.

single driver performs well at midrange and still fulfills the exacting demands of the very high frequencies, all without distortion at high power. A photo of the usual inside front view of the 2-way suspension system is shown in Fig. 8-5. Although layouts may vary widely, according to the sizes and design requirements, the 3-way is shown roughly in Fig. 8-6. The acoustic suspension speaker system has enjoyed a well-deserved popularity for more than a decade with its remarkable performance at moderate prices. Such features as wide frequency range, low distortion, and unexcelled smoothness across the audible range have been available from the beginning and all improvements in design over the years have been compared against that standard. The 2-way speaker is pictured in Fig. 8-7.

Shortcut to the Circle of Sound

The use of omnidirectional speakers suspended from the ceiling will provide a novel feeling of sound from all directions

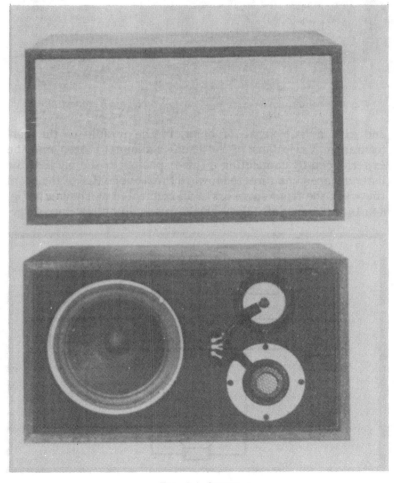

Fig. 8-6. 3-way.

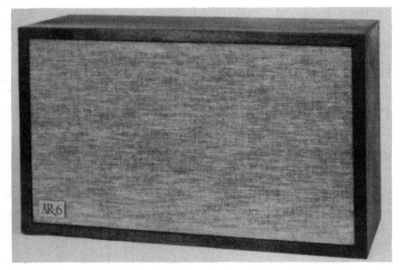

Fig. 8-7. AR-6 speaker system. (Courtesy Acoustic Research, Inc.)

and give a true sense of being in the middle of the performance. Variations in the usual 2-channel stereo may be experienced by connecting a third speaker in the back of the listening area and directed toward the front center of the room (between the front speakers) and connected as shown in Fig. 8-8. In this case, all three speakers should be similar, direc-

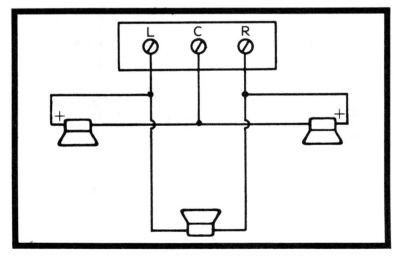

Fig. 8-8. Three-speaker stereo.

tional units projecting the sound forward toward the listener. Another arrangement of speakers to enjoy a circular or "surround effect" is displayed in Fig. 8-9 using the same amplifier connection. The actual construction of speakers that project sound in all directions may be undertaken by installing a 4" or 5" speaker in each panel of a cube-shaped cabinet. Directions are normally supplied with such kits, including instructions for covering with a coarse-weave cloth. All speakers in each cube are connected to the same source or channel, polarity and impedance must be observed. The enclosure must not vibrate, and if any problem arises here, it will have to be corrected by using more screws or braces. Although building a speaker system may involve considerable work, depending on your ability along these lines, much pleasure as well as savings could result. Then, there is the chance to meet a specific requirement or need to fit your individual conditions that may be satisfied only by building your own. The easy way is to buy high fidelity speakers unless the pre-packaged system is too far from filling your specifications. Building your own speaker system offers a wide selection of sizes as well as levels of performance, and the convenient installation in cabinets, walls, ceilings, and many existing enclosures. It is easier to update your system as your budget permits.

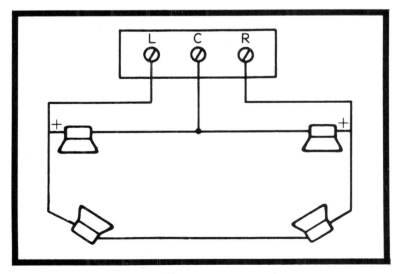

Fig. 8-9. Four-speaker sound.

UNIQUE LAYOUTS FOR DEPTH EFFECT

The larger speaker may or may not improve bass response, it depends on the size of the enclosure and other conditions as well. Locating a speaker where walls join, such as the corner of a room at floor level or ceiling level, will often improve the low frequency output. While your system is operating, try the floor speaker across a corner and see, then move it out a foot or so while noting the change. This may only be checked by trial, so move the unit around until the best result is discovered. Additional low frequency response resulting from exact placement of speakers may reach surprising proportions under most favorable conditions.

A new shape in sound is offered by Utah in their Celesta Series speakers with a chassis cast under extremely high pressure, which assures perfect alignment of critical moving parts. This functional, high styling is the key to provide large open areas for freedom of cone movement, taking full advantage of the latest RF heating techniques and new formulations of epoxy to guarantee an indestructible bond between the barium ferrite magnet and a massive steel return circuit. An exclusive internal dust seal protects the precision voice coil gap from "stray whiskers" that could work loose from the magnet during shipment. Shallow profile with 8-inch models extending only 3⅝" deep, and 12-inch models a mere 5¾" deep. The many features include the large windows in the frame, finger-tip terminals with space-age Teflon insulation, baked on lacquer finish, and six models available to fit each specific requirement. The speaker terminals are even color-coded with RED for positive phasing. The C8JC3 and C12PC-HF are 3-way, coax models with cloth roll suspension of the woofer cone, and this improved suspension is also offered on the 2-way C8JC2A and C12PC-2A models. The C12PC-HF tweeter is a new type compression horn. Both 3-way models boast internally mounted crossover capacitors plus prewired tweeter controls. The top of the Celesta Series is the C12PC-HF which is a 3-way coax with a peak audio capacity of 45 watts, a 20-ounce magnet, a 1½" voice coil, and a frequency response of 25 through 20,000 Hz. Outside diameter is 12¼". It is priced at $34.95, and is pictured in Fig. 8-10.

The construction of the Utah speaker is shown in the cutaway view in Fig. 8-11 along with the individual parts involved. This makes it easy to see what happens when the voice coil or spider strikes a metal part of the speaker to cause bottoming. When the speaker is mounted in a sealed enclosure, the cone travel is reduced for any given applied power; and, for this reason, a higher power level may be applied without damage under these conditions. In the case of an electric organ, power ratings must be reduced, possibly as much as 50 percent, as the organ tones approach square waves which amount to sustained peaks rather than the usual short duration peaks.

Fig. 8-10. Utah model C12PC-HF 3-way speaker. (Courtesy Utah Electronics.)

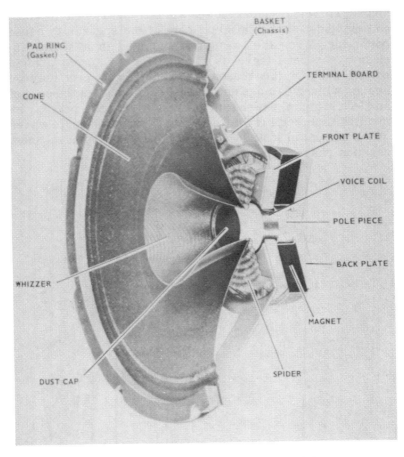

Fig. 8-11. Speaker construction & nomenclature.

CONSOLE CABINETS

Altec model A7-500W-II **Voice of the Theatre** system features a hand-finished, sculptured walnut cabinet patterned in natural wood grain with a delicate curved-wood fretwork grille. The new Symbiotik Driver is used in the system which will handle twice the power of previous designs along with a new crossover that features an attractive cast-aluminum escutcheon and is rated at 100 watts. The voice coil wire is made of flat aluminum ribbon that has been edge-wound for greater efficiency, and that withstands temperatures as high as 350 degrees Fahrenheit. Frequency response is 30-20,000 Hz with a power rating of 50 watts. The model may be used with amplifiers expressing an equal continuous power rating. Specifications include: impedance 8-16 ohms, crossover

Fig. 8-12. Altec model A7-500W-II speaker system. (Courtesy Altec Lansing.)

Fig. 8-13. Altec Milano stereo ensemble (885-871). (Courtesy Altec Lansing.)

frequency 500 Hz, and dimensions of 44" H, 32" W, 25" D. Horn coverage of the entire audio spectrum is offered with a cast-aluminum sectoral horn for mids and highs and a wooden horn for the low frequencies. This controls wide angle distribution of sound at all frequencies and the natural uncolored sound required by professionals. This model is priced at $555, and pictured in Fig. 8-12.

The Altec Milano Stereo Ensemble (885-871) reflects the cabinet-maker's art with the 885A equipment cabinet of pecan-finished hickory handcrafted in dramatic, deeply sculptured design to match the 871A Milano Speaker Systems. Ample space is provided for tuner, amplifier, tape deck, record player, and storage of records, tapes, and miscellaneous. Overall dimensions 8'3" wide, 29¼" high, and 19¼" deep. The 817A speakers have a frequency response of 35-20,000 Hz with a power rating of 50 watts and may be used with amplifiers having up to equal continuous power rating. Impedance is 8 ohms, and the crossover frequency 800 Hz. The ensemble is pictured in Fig. 8-13.

An additional ensemble is the Flamenco 2-section equipment cabinet in handcrafted heavy oak, and providing ample space for stereo components such as receiver, record player and tape recorder. Dimensions are: 8'3½" overall, 27¾" high, and 19⅜" deep. Also available in matching 3-section equipment cabinet (884-848) with an overall width of 9'11", height 27¾", depth 19⅜". The speaker systems have the unbounded dynamic range normally associated only with theatre systems in smoothness, flatness of response, high efficiency, and body without coloration. Specifications are: frequency response 35-20,000 Hz with a power rating of 50 watts and an impedance of 8-16 ohms utilizing a crossover at 800 Hz. The speaker components include a massive sectoral horn of heavy cast-aluminum for wide angle distribution of all frequencies from 800 Hz to beyond audibility.

APERIODIC DESIGN

The aperiodic loudspeaker system represents a design without a fixed resonant frequency or repetitive characteristic but optimizes the efficiency of power transfer between amplifier and speaker. The end result of such design is not only better bass and transient response, but an alleviation of the

Fig. 8-14. A-25 aperiodic speaker system. (Courtesy Dynaco, Inc.)

Fig. 8-15. A-10 aperiodic speaker system. (Courtesy Dynaco Inc.)

problems which result in Doppler distortion as well. Vastly improved vocal expression is evident, and women in particular will note that this performance has been achieved without the offensive stridency or harshness which accompanies exaggerated midrange or tweeter characteristics. The aperiodic feature is obtained in this case by using a highly damped vent rather than a reflex port which is carefully controlled by filling a narrow slot with a critical density of fiberglass. Each speaker system is adjusted by observing the back EMF of a 5 Hz square wave which has been fed into the speaker system. Damping material is added and compressed until an optimum square wave is shown on the oscilloscope. Although no speaker system is designed to reproduce 5 Hz square waves, it is important that overshoot on such long excursions be minimized since the harmonics generated by the overshoot do fall in the normal audio spectrum. The reduction of such spurious responses yields the precise bass desired.

Dynaco's A-25 speaker system as pictured in Fig. 8-14 was introduced to fill the need for a system under $100 ($79.95) that fulfilled most listening requirements. The RMS power rating of the A-25 is 20 to 60 watts per channel with a 1500 Hz crossover. The smaller A-10 model as pictured in Fig. 8-15 utilizes a 6.5" woofer, has a 2500 Hz crossover and the same output handling capability and same 8-ohm impedance as the A-25. The A-10 is priced at $99.95 per pair.

Overall System Checks and Summary

Chapter 9

A unique though effective way of checking out your system is by simply walking around your main listening area and noting exactly what you hear. Borrowed from security system technicians, this method comes closer to an accurate evaluation than anything except analysis by an expensive, fully-equipped laboratory. The end result could hardly be more satisfying. If you like what you hear, and the way that you hear it, just stand pat. But if something is not quite right, like too much tin and not enough crisp highs, do something about it. No matter what the problem, you do not have to upset that budget for months to correct it, as there are many ways to bypass minor faults...even simple but careful adjustments may suffice. Refer to manufacturer's literature on your equipment. Recommendations as to specific procedures will be helpful to non-technical users, and will eliminate chancing accidental damage to some units as well.

SIMPLE SHORTCUTS AND ADJUSTMENTS

Before considering advantageous adjustments, make sure of proper placement. Are we avoiding proper ventilation for the sake of convenience, and if so, correct that fault first of all. True, heat from the solid-state AM-FM stereo receiver-amplifier is but a small fraction of that generated from the older vacuum tube units, but adequate ventilation must still be provided for reliable, long-lasting performance. Do not shove that receiver-amplifier against the wall or the back of a bookshelf, leave two or three inches at least to permit air circulation which ensures that built-in solid-state reliability. Sometimes it is possible to achieve better reception from your FM receiver when tuning by ear, with AFC off, listen carefully for the center of the valley between the hiss. As you tune across the louder hiss on either side of the station, the hiss level drops and disappears at the exact center, rising again as tuning continues up the other side. Minimum or no hissing noise is the correct peak, and this may be missed when using the meter or eye tube in some sets due to slower or less accurate response. Try it on your receiver, using a weaker station and not paying attention to music, only the background noise which will reach a sharp null or zero point.

Speakers with forward sound patterns as usually found in conventional speakers are often improved when placed across the corner of a room. This may be well worth a try if you are using this type of unit: move them out from the corner a little at a time until the best spot is apparent.

If a little more pickup is desired on the FM receiver, and a regular outside antenna is not available, lay a piece of twin-lead on the television lead-in by tying a piece of twine around the two and connecting the other end to the FM antenna terminals. Although no electrical contact is made, FM signals will be picked up from the TV lead wire. The TV will not be affected, but of course the FM received will not be nearly as good as a special FM antenna would deliver as the TV antenna elements are not the correct length for FM.

A surprising number of audio problems turn out to be simple things instead of the complicated one we so often anticipate. Dirty plugs which are actually slightly corroded may be causing any type of intermittent fluctuation in volume level or even distortion, and other examples of average troubles may lie in broken cables or leads, cold solder connections, corroded switches or controls, and even incorrect connections if something has been added or changed recently. In case of failure these ordinary things should be first checked before looking for anything more serious.

STEREO ACCESSORIES

Dynaco model PAT-4 preamplifier offers extremely wide frequency response to avoid any possible effect on square waves or other signals being passed which require wide-band, low phase shift response for accurate reproduction. The preamp may be used with virtually any power amplifier whether it be tube or transistor. Performance is characterized by a remarkably low level of noise and distortion, in fact below that which could be measured with commercial test equipment, as a result of the carefully selected silicon transistors in

Fig. 9-1. Dynaco model PAT-4 stereo preamplifier. (Courtesy Dynaco Inc.)

a circuit design utilizing both DC and AC feedback. Specifications are: frequency response, high level inputs plus or minus 0.5 db from 10 Hz to 100 kHz, low level inputs plus or minus 1 db from 20 Hz to 20 kHz (equalized); distortion at rated 2-volt input, THD less than 0.05 percent with any combination of test frequencies; hum and noise, magnetic phono 70 db below a 10 mV signal input, high level 85 db below a 0.5 volt input signal; gain, magnetic phono 54 db at 1,000 Hz (3 mV for 1.5V out), high level 20 db (0.15 V for 1.5V out); impedance, magnetic phone 47K, tape head 100K, high level 100K, audio output 600 ohms, amplifier input 10K or higher nominal load. Referring to the photo of the PAT-4 as illustrated in Fig. 9-1, four bass and treble controls use independent concentric knobs in a tone control circuit providing smooth, continuous adjustment with a true center-flat position without requiring disabling switches. The balance control offers critical adjustment of small differences through initial 90 degrees from center, although allowing complete cancellation of either channel at its extremes. A rotary high frequency filter switch has a flat position plus three half-octave steps, and the output jack at 600 ohms for headphones or tape recorder provides more gain than the conventional back panel tape recorder output while enabling full use of all controls during recording; front panel stereo input for high level source such as tape recorder or guitar amplifier. Mono input can be combined with other mono program material from another source on the other channel if desired. Power consumption is 5 watts, 120-240V AC 50-60 Hz. Power switch is illuminated. The unit is available in kit form at $89.95 or assembled and tested at $159.95.

The Pioneer model SX-9000 AM-FM stereo receiver features a built-in reverberation unit and is pictured with the bottom controls exposed and with cover closed in Fig. 9-2.

The diamond stylus timer (Park PNT1 at $5.25) actually keeps track of wear and indicates point condition before damage can result to your recordings. Internal magnet of timer senses turntable revolutions to 250 hours, then may be reset quickly after checking.

Test records and tapes available include: Audiotex 30-200 (record, $4.98); Realistic 50-1001 (record, $2.49); Audiotex 30-214 (reel $6.80); Audiotex 30-212 (cassette $5.69); Audiotex 30-213 (cartridge $7.98).

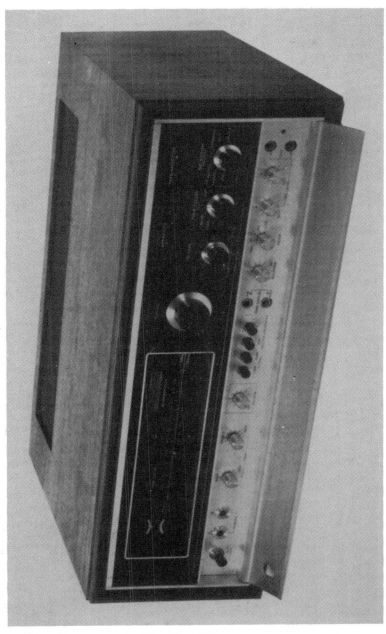

Fig. 9-2. Pioneer model SX-9000 AM-FM stereo receiver with reverberation. (Courtesy U.S. Pioneer Electronics Corp.)

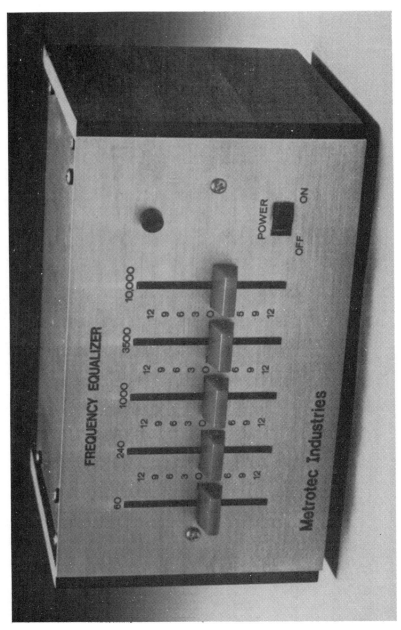

Fig. 9-3. Metrotec model FEW-1 frequency equalizer. (Courtesy Metrotec Electronics, Inc.)

Wireless Stereo Speaker System ($150), is an old idea improved by updating. The use of FM affords far better quality than the AM method. Connect hi-fi output to encoder box for transmission of two channel stereo throughout house wiring to left and right speakers plugged in anywhere. Each speaker has its own decoder and amplifier (either left or right channel) enclosed. Exceptionally handy for patios, garages, etc.; simply plug into the AC receptacle for high fidelity music from your own stereo any place where a receptacle is available. Even stretch a little with an ordinary power line extension cord if you wish.

Hi-FI Sentinel (Realistic 42-2932 at $5.95) includes duplicate AC sockets and switch in metal enclosure to connect to record player. Small lamp may be plugged in other socket, so that when the changer stops, the amplifier and lamp will go off as well.

Stereo Headphone Control Box (Lafayette 99-00242 at $4.95) permits individual left and right earphone adjustment from easy chair. The 3-position slide switch connects phones to left or right channel as desired.

Metrotec model FEW-1 frequency equalizer is a 2-channel direct coupled four-stage amplifier with slightly less than unity gain. The emitter follower input ensures high input impedance and a two-stage differential amplifier with extremely high feedback follows. The circuit compares input to feedback and allows 24 db range of control. The output stage design provides very low output impedance with high output voltage and extremely low distortion. This allows its use with any existing amplifier or receiver. This unit gain amplifier may be inserted into your stereo system between preamp and the power amplifier or at the tape monitor jacks on your receiver. It has five manually operated controls that vary segments of the audio spectrum at 60, 250, 1,000, 3,500, and 10,000 Hz; this is through a range of plus or minus 12 db. The unit is available at $99.95 wired and tested or about 20 percent cheaper in kit form. (See Fig. 9-3.)

DEFINITIONS PROVE THE POINT

In judging the various claims enunciated by equipment sales personnel and checking through the specifications listed, you may find the real proof you are looking for to make the final decision on a particular piece of stereo equipment. Definitions often drive the point home, and as these are proven, there can be little need for further anxiety.

HOW TO EVALUATE WHAT YOU HEAR

Anyone currently owning a stereo system can check the merits of the new sound in his own home for very little before going further into his budget. No second stereo amplifier is required—just the adapter or converter and two more speakers. The front speakers should be identical and in your system. The rear speakers which you must pick up should be almost as good and matched at 8 ohms. The smaller the variation in impedance, the better your test will be and the more accurate too. You may find that something really new in sound has been revealed to you...and you may like it.

Glossary of Hi-Fi Terms

A

A-B Test - Comparison method by switching from one system to another while maintaining proper operation for audio evaluation by ear.

Acoustic Feedback - The howling sound caused by sound vibrations from a speaker being picked up by the adjacent microphone, turntable, or other transducer.

Acoustic Filter - A device for suppressing certain frequencies in the audio range by absorption.

Acoustic Impedance - Load or opposition which a device offers to sound waves. It is measured in ohms and may vary with frequency. Also referred to as impedance, as in transfer of energy.

Alignment Suspension - A design involving suspension of a speaker cone in an airtight or sealed enclosure to enhance the low frequency response of the loudspeaker.

Alignment - The adjustment or arrangement of units in a system for best results—in a tuner, proper adjustment of tuned circuits; in the tape player, relation of head to tape; in record player, tone arm to record groove positioning.

Alnico - A metal alloy offering exceptional magnetic qualities and retention of these properties. Used in loudspeaker magnets.

Ambience - Surround or encircling as applied to sound fields in four-channel or even modifications of two-channel systems.

Amplifier - A unit or part of the system that enlarges the small signal from the source (FM tuner, preamp, tape, phono, microphone) to a size sufficient to drive the loudspeakers; may be part of the AM-FM receiver or a separate unit, as a power amplifier.

Anti-skating - A method to prevent skidding of tone arm by centering the stylus in the recording groove with automatic balance. This action reduces record wear.

Audio Frequency - The range of frequencies to which the human ear responds, covering an audible range of sound waves from 15 Hz to 20 kHz (20,000 Hz). Usually abbreviated AF and important to all hi-fi or stereo listeners.

Audiophile - An enthusiast or person having an ardent interest in perfect sound reproduction through modern techniques, and in the ultimate in personal listening pleasure.

B

Baffle - The mounting board, support, or shielding wall which increases speaker loading and eliminates sound interference from the rear of the loudspeaker.

Balance - A blending or equalizing of channel sound outputs to effect their realistic proportions.

Bass Reflex - A method of improving bass performance of a loudspeaker by allowing low frequency sound from the back to emerge through an opening in the cabinet or enclosure of the speaker.

C

Capstan - The shaft to the motor that drives the tape past the heads on a tape machine, and which controls the speed according to the diameter of the capstan.

Cartridge - The pickup device on the end of the phono tone arm converting stylus motion on the record grooves to electrical waves to the amplifier. The endless loop of tape in a plastic magazine for automatic tape recorders or players.

Cassette - An automatic tape system for recording or playback with $\frac{1}{8}$" magnetic tape between two tape reels having stops on each end and enclosed in a plastic magazine to be driven by the capstan in either direction.

Channel - A path for reproducing sound, from mechanical to electrical or electrical to mechanical according to the function.

Compact - A streamlined hi-fi system, usually including a record player with built-in amplifiers and two detachable speakers for proper stereo setup.

Compatibility - The ability of different systems or equipment to work together harmoniously and at least to a degree satisfactorily.

Compliance - The ability to flex readily, as of a stylus to yield to recording groove variations or a loudspeaker cone to yield to the magnetic field fluctuations.

Component - An essential part of a hi-fi or stereo system, i.e., tuner, receiver, amplifier, turntable, speaker, tape deck, microphone, or any accessory or attachment.

cps - Formerly referred to number of cycles per second. The term has been replaced by hertz (Hz).

Crossover Network - A device for dividing a signal according to frequencies. The outputs are fed to appropriate speakers, such as high frequencies to tweeter and low frequencies to woofer.

Crosstalk - Signal originating in one channel which finds its way into another channel.

Cueing - A method to permit the tone arm to be raised or lowered from record surface without scratching.

D

Damping - Reduction of energy by a loading effect to eliminate oscillation, vibration, or sharpness of response.

Decibel - Abbreviated db, the standard unit for indicating relative power, sound, voltage, or current levels. The measurement is not linear: 3 db is double 10 db equals ten times, and 17 db is fifty times the power.

Decoder - A device for recovering coded information, such as for obtaining four-channel information or obtaining second channel information in FM stereo.

Derived - Four-channel sound from two-channel material with adapter and two more speakers; center-channel sound derived from two-channel and fed to an extra speaker.

Discrete - True, individual four-channel reproduction from four distinct and separate signals as recorded on four separate tracks, played back on four-channel tape machine and fed into four-channel amplifier with four speakers.

Distortion - Any difference between the output and the original input other than in amplitude. It is frequently expressed in percentage and divided into types. Harmonic distortion alters the relationship between tones, IM distortion causes new tones by mixing original tones.

Dolby System - A method of reducing noise by boosting relative signal level of low passages while leaving loud sections of music as they are when recording. On playback, low passages are pushed back to normal but noise remains reduced. The system shows a typical noise reduction of 10 db or more.

Dynamic Speaker - A moving-coil speaker where the coil is attached to the cone or diaphragm and is electrically connected to the output of the system amplifier. The coil floats in the permanent magnetic field with the current through the coil causing a movement back and forth according to its variations. This results in the conversion of the electrical sound energy into mechanical sound energy through the attached diaphragm.

E

Efficiency - Usually as applied to loudspeakers a percentage expressed by the power in divided by power out.

Encoding - A process for converting to coded form without disturbing the original but at the same time adding information. Encoded four-channel may be played on two-channel equipment. With decoder, the complete four-channel material is reproduced.

Equalization - Matching frequency response of records and tapes for even amplifier output by means of compensating networks.

Equalizer - An adjustable RLC device to insure constant response to all audio frequencies.

Erasing Head - A device for clearing or preparing the magnetic tape for recording on a tape player-recorder.

Extended Play - A seven inch (45 rpm) recording usually offering about 50 percent additional playing time.

F

Fade Out - A smooth drop off in volume at the end of a recorded tape or disc.

FET - A field effect transistor offering major advantages of a vacuum tube along with those of a solid-state device.

Flutter - Frequency changes caused by high speed deviations in turntable or tape feed.

Flywheel Tuning - A heavy wheel on the main-tuning shaft to provide more momentum and ensure smooth operation of the dial.

FM Stereo - The transmission of two signals in a single transmitting channel, also called multiplex.

Four-Track Tape - Magnetic ¼" tape with four recording tracks for two-channel stereo, two tracks in one direction and other two opposite direction. Four-channel stereo uses all four tracks in same direction.

Frequency - The measure of sound vibrations or electrical alternations per second as expressed in hertz (Hz). The term hertz has replaced the formerly used terms cycles or cycles per second.

Frequency Distortion - The unequal amplification or transfer of audio frequencies resulting in a loss in quality of reproduction.

Frequency Drift - An undesirable effect in FM receivers, usually eliminated or corrected by AFC circuitry.

Frequency Response - The scope over which the equipment may perform effectively within stated limits, expressed in hertz (Hz).

G

Gain - The degree of amplification between points in a system. Gain may be varied by the volume or gain control, or controlled automatically at a preset level by the AVC circuit. Usually gain is measured in decibels (db).

H

Harmonic Distortion - A condition resulting from harmonics added during amplification which alters the relationship between tones.

Headphone - A small sound reproducer for private listening attached to a headband. Usually they are provided in pairs with one for each ear. Room noise is shut out and stereo separation is improved.

Hertz - A measure of frequency usually referring to vibrations or alternations per second. Abbreviated Hz, the hertz replaces cycles or cycles per second in this connotation.

Hi-Fi - The abbreviated and popular form of the term **high-fidelity** which refers to a system capable of reproducing music as originally played. Stereo systems are expected to qualify as hi-fi.

Highs - Indicates high-pitched sounds such as those near the top audible limits. They are emphasized by the tweeter speaker.

Hum - A low-pitched noise usually introduced into the amplifier from its source of power and amplifier along with the music. Several possibilities exist, but correction is normally technical in nature.

Hysteresis Motor - A type of synchronous motor without prominent poles and DC excitation. Starting results from hysteresis losses induced in the steel secondary by a revolving field. Frequently used with more expensive record players to ensure constant speed at all times.

I

Idler - The small rubber wheel that contacts the inside rim of the phonograph turntable to drive it at a preset speed.

IHF - Standards set up for the guidance of its members by the Institute of High-Fidelity Manufacturers.

Impedance - A measure of opposition to the transfer of energy from one unit in the system to another. It is expressed in ohms and varies with frequency.

Integrated - A unit of a hi-fi system formed by the combining of two or more sections on a single chassis; as a receiver which could include the tuner, preamp, and main amplifier. An integrated circuit (IC) is a solid-state device comprising several component parts (transistors, diodes, resistors) internally connected to form a complete electronic circuit.

Intermodulation Distortion - Distortion caused by new tones which result from mixing of sums and differences of original tones. Abbreviated as IM.

ips - Inches per second: a measure of tape speed. Common standard speeds are 1⅞ ips, 3¾ ips, 7½ ips, and 15 ips.

L

Labyrinth - A specially constructed loudspeaker enclosure to prevent acoustic standing waves.

Localization - Positioning of sound image. For two-channel stereo the image is in front of the listener between the two speakers. Four-channel centers the sound image among all four speakers to have the listener surrounded by sound.

Loudness Control - Control for reducing overall volume level.

Lows - The low pitches or bass sounds that are handled by the larger, heavier speaker known as the woofer in the system.

M

Matrix - A method of combining four channels into two channels as rear channel information is encoded on regular front channels.

Mid-Range - The frequency range between bass and treble often handled by a mid-range horn operating at about 400 to 3,000 Hz.

Monophonic - Mono or single channel sound, formerly called monaural, like AM radio which uses one speaker operation.
Multiplex - FM transmission of two or more signals on a single channel as in stereo.

N
Noise - All unwanted sounds such as hum, hiss, crackle, popping, and all forms of distortion.

O
Octave - The eighth full tone above any given tone. It has twice as many vibrations per second, as the given tone. Also the distance between two frequencies which have a ratio of two to one.
Omnidirectional - A type of speaker which disperses sound all around equally in a 360-degree pattern. No direction is favored over any other.

P
Phase - The relationship of sound or electrical waves to each other from a standpoint of time.
Phase Distortion - A phase shift lacking in linearity which upsets the usual timing sequence between tone and related overtones.
Pickup - The unit or cartridge on a phonograph tone arm that translates the motion of the stylus in the record grooves into electrical energy to be amplified.
Power Bandwidth - The width of an amplifier's undistorted power output, outside of which it falls below 50 percent of its rating. IHF standards are used.
Preamplifier - A device designed to boost extremely weak source signals to a sufficient level to drive or properly operate the main amplifiers. Usually includes equalizing and tone control circuits, may be part of the receiver.
Presence - Natural feeling of reality as though the musicians are actually present, as imparted by a stereo system, the degree may be increased with four-channel sound.

Q
Quadraphonic - Four-channel stereo, the "surround sound" of four-speaker reproduction.

R
Receiver - A unit of the hi-fi system, usually includes AM-FM tuner, preamp, and power amplifier.

Reverberation - The continuation of sound resulting from reflection of sound waves off exposed surfaces in the area. Provides a dimensional satisfaction characteristic of a live performance, usually more pronounced in four-channel information.

RMS Power - Amplifier's continuous power output expressed in watts per channel, usually specifying ohms and frequency. The abbreviation RMS refers to Root Mean Square or effective rate.

Rolloff - The sudden or appreciable drop in the amplitude versus frequency curve as frequency limits are approximated.

Rumble - Low frequency vibrations in phono or tape player which are superimposed in the output.

S

Separation - The amount channels are kept apart which affects the true stereo quality. Expressed in decibels with 35 db or more being desirable, and normally called "stereo separation."

S-N Ratio - The signal to noise ratio expressed in db compares the level of the desired sound versus the level of all undesired sounds, and should exceed 45 db.

Speaker System - A combination of speakers in a single enclosure including crossover network plus tweeter and woofer in two-way, or with tweeter, mid-range horn, and woofer in three-way system. Other speaker systems include additional speakers for specific purposes, often to achieve certain directional advantages.

Stereo - A contraction of stereophonic which is a sound reproduction system to preserve the realistic quality of the original by eliminating the single speaker source. Stereo recordings must be made with at least two separate microphones on separate tracks. The stereo equipment requires two separate amplifiers with separate speakers for reproduction, and although the amplifiers are usually combined on a single chassis or board, the speakers must be separated for the true stereo effect.

Stylus - The part of the phono cartridge that contacts the record groove to carry the recorded information to the cartridge for conversion to a corresponding electrical signal. Diamond tipping is desirable for quality reproduction.

Subcarrier - The second harmonic of the control signal used in FM stereo broadcasting.

Synthesizing - A method of producing four-channel effect from conventional two-channel source, usually by adding synthetic material for rear channels as provided in various ways by matrix decoders.

T

Tape Deck - The unit containing mechanism for open reel, cartridge, or cassette tape recording and playback. Although a preamp may be included, only motorboard is normally provided.

Tape Monitor - A third tape head is included on some recorders to allow instant checking as recording is being made. This head monitors the tape a moment after it passes the record head.

Tone Control - A variable control to increase or decrease the low or high frequency response as desired.

Transient Response - The response of a system to a momentary change of input. The resulting distortion in amplifiers may be minimized by designing the circuitry to have a short transient response time.

Tuner - The radio receiving (AM and FM) part of the system, and currently popular along with the preamplifier and power amplifier to form the AM-FM stereo receiver.

Tweeter - The high-frequency speaker, usually reproducing those sounds above 3,000 Hz.

V

VU Meter - A meter for indicating "Volume Units" to assist monitoring of gain level during tape recording and playback.

W

Woofer - The low frequency speaker, specializing in the bass part of the audible spectrum.

Wow - A low-frequency distortion in the sound resulting from slow changes in speed of tape or record player.

This author expresses thanks to the following members of the audio fraternity for their unselfish assistance in supplying information and photographs on short notice:

Acoustic Research, Inc.
Benjamin Electronic Sound Corp.
British Industries Company
BSR McDonald
Dynaco Inc.
Electro-Voice, Inc.
Elpa Marketing Industries, Inc.
Empire Scientific Corp.
Fairfax Industries
Fisher Radio
Harman-Kardon, Inc.
Kenwood
KLH Research & Development
Marantz Company, Inc.
Metrotec Electronics, Inc.
Revox Corporation
Sansui Electronics Corp.
Sherwood Electronic Laboratories, Inc.
Shure Brothers, Inc.
Sony Corp. of America
Toshiba America, Inc.
United Audio Products, Inc.
U.S. Pioneer Electronics Corp.
Utah Electronics, also local dealers who generously gave of their time, including Lafayette Radio Electronics.

Index

A

AC operation, battery,	120
Accessibility and control features,	119
Accessories,	
—connecting,	138
—stereo,	204
Action, amplifier,	14
Additional features in other units,	74
Additional speaker systems,	173
Adjustments, observe tracking, anti-skating and other	86
Air suspension speaker system,	163
Allowing a safety margin,	167
AM-FM receiver,	
—direct recording from,	123
—evaluating the,	17
AM-FM receivers and amplifiers,	25
Amplifier action,	14
Amplifiers, FM,	25
Amplifier section, the stereo	36
Answers to expect,	13
Anti-skating, and other adjustments, observe tracking,	86
Aperiodic design,	199
Approach, the kit,	28
Arrangements, suspension,	189
Ask, questions to,	13
Assembling speaker cabinets for unusual balance	188

B

Basic terminal layouts,	143
Battery-AC operation,	120

C

Cabinets, console,	196
Cables and connectors,	137
Cables, importance of connecting,	140
Capability, power handling,	166
Cartridge, trackability,	93
Cartridge (8-track), cassette, reel-to-reel, or,	117
Cassette, reel-to-reel, or cartridge (8-track)	117
Channel sound, four,	60
Checks and summary, overall system,	203
Chromium tape, use of	136
Circle of sound, shortcut to the,	191
Cleaning and degaussing, simple	121
Closet enclosures,	184
Connecting accessories,	138
Connecting cables, importance of,	140
Connection procedure,	65
Connectors and cables,	137
Considerations,	59
Considerations, major, in record players,	8
Console cabinets,	196
Compact systems,	87

221

Constant voltage lines,	152
Control features, accessibility and,	119
Controls,	
—and output requirements, flexibility,	36
—exterior level,	146
—proper use of external,	170
Converter, updating your stereo with a,	61
Cost considerations,	59

D

Definitions prove the point,	209
Degaussing and cleaning, simple,	121
Depth effect, unique layouts for,	194
Design, aperiodic,	199
Developments, new record playing system,	112
Direct recording from AM-FM receivers,	123
Drive, importance of mechanism,	115

E

Economy units,	87
Effect on what you hear, power and its,	168
Enclosures,	
—closet,	184
—wall,	184
—shelf,	184
—speaker,	183
Evaluate what you hear, how to,	210
Evaluating the AM-FM receiver,	17
Exceed minimum performance,	19
Expect, answers to,	13
External controls, proper use of,	170
External level controls,	146
Extra speakers, installing,	144
Extras, higher priced,	28

F

Features,	
—accessibility and control,	119
—in other units, additional,	74
—private listening,	147
—tape player,	10
—turntable,	96
Flexibility, controls, and output requirements,	36
Four channel sound,	60
Four-channel AM-FM stereo receiver, wireless remote,	84
Frequency range, importance of,	171

G

Getting maximum output to speaker systems,	141
Getting more "lows" for less money,	169
Glossary of Hi-Fi terms,	211

H

Handling capability, power,	166
Headphones, plugging in stereo,	146
Higher-priced extras,	28
Hi-Fi,	
—speaker systems,	155
—terms, glossary of,	211
How to evaluate what you hear,	210

I

Importance of,	
—connecting cables,	140
—frequency range,	171
—mechanism drive,	115
Improving your system with low-cost accessories,	20
Installing extra speakers,	144

J

Jacks and plugs, 151

K

Kit approach, the, 28

L

Layout for depth effect, unique, 194
Layouts, basic terminal, 143
Level controls, external, 146
Level meters, 119
Lines constant voltage, 152
Listening features, private, 147
Low-cost accessories, improving your system, 20
"Lows" for less money, getting more, 169

M

Major considerations in record players, 8
Margin, allowing a safety, 167
Mechanism drive, importance of, 115
Meters, level, 119
Microphone operation, 17
Microphones, types of, 17
Money, getting more "lows" for less, 169
More shortcuts to Quadraphonic sound, 69

N

New record playing system developments, 112

O

Observe tracking, anti-skating and other adjustments, 86
Operation, 68
—battery-AC, 120

—microphone, 17
—speaker, 24
Output requirements, flexibility, controls, and 36
Output to speaker systems, getting maximum, 141
Overall system checks and summary, 203

P

Performance,
—exceeding minimum, 19
—requirements for reliable, 85
Players, stereo record, 85
Plugging in stereo headphones, 146
Plugs and jacks, 151
Power,
—and its effect on what you hear, 168
Power handling capability, 166
Power requirements, 59
Private listening features, 147
Procedure, connection, 65
Program sources 7
Proper use of external controls, 170
Pushbutton value, 122

Q

Quadraphonic sound, more shortcuts to, 69
Questions to ask, 13

R

Range, importance of frequency, 171
Receivers, AM-FM 25
Record players,
—stereo, 85
—major considerations in, 8
Record playing system developments, new, 112
Recorders, tape, 114
Recording from AM-FM receiver, direct, 123
Reel-to-reel, or cartridge (8-track) cassette, 117

Reflector speaker system,	162
Reliable performance requirements,	85
Requirements for reliable performance	85
Requirements, power,	59

S

Safety margin, allowing a,	167
Shelf enclosures	184
Shortcut to the circle of sound,	191
Simple cleaning and degaussing,	121
Sound,	
—four channel,	60
—quadraphonic, more shortcuts to,	69
—shortcut to the circle of,	64
Sources, program,	7
Speaker cabinets for unusual balance, assembling,	188
Speaker enclosures,	183
Speaker operation,	24
Speakers, installing extra,	144
Speaker system,	
—reflector,	162
—air suspension	163
Speaker systems,	
—additional,	173
—getting maximum output to,	141
—Hi-Fi	155
Stereo,	
—amplifier section, the	36
Stereo accessories,	204
Stereo headphones, plugging in,	146
Stereo with a converter, updating your,	61
Stereo receiver, wireless remote 4-channel AM-FM stereo receiver,	84
Stereo record players,	85
Suspension arrangements,	189
Suspension system, speaker, air,	163

System,	
—checks and summary, overall,	203
—reflector speaker,	162
—wiring,	65
Systems,	
—additional speaker,	173
—compact,	87
—Hi-Fi speaker,	155

U

Unique layouts for depth effect,	194
Units,	
—economy,	87
—other, additional features in,	74
Unusual balance, assembling speaker cabinets for,	188
Updating your stereo with a converter,	61
Use of chromium tape,	136

T

Tape,	
—player features,	10
—recorders,	114
—use of chromium,	136
Terminal layouts, basic,	143
Terms, glossary of Hi-Fi,	211
Trackability, cartridge,	93
Turntable features,	96
Types of microphones,	17

V

Value, pushbutton,	122
Voltage lines, constant,	152

W

Wall enclosures,	184
What you hear,	
—how to evaluate,	210
—power and its effect on,	168
Wireless remote 4-channel AM-FM stereo receiver,	84
Wiring system,	65